THE UNEXPLAINED

UFOs • Ghosts • Time Warps • ESP

MARSHALL CAVENDISH

This edition published by
Marshall Cavendish Books
(a division of Marshall Cavendish Partworks Ltd)
119 Wardour Street, London W1V 3TD

Copyright © Marshall Cavendish 1994

First printing 1994
This printing 1994

ISBN 1 85435 722 0

All rights reserved. No part of this publication may be reproduced, stored in a retrieval system or transmitted in any form or by any means electronic, mechanical, photocopying, recording or otherwise, without the prior written permission of the publishers and the copyright holder.

Some of this material has previously appeared in the Marshall Cavendish partwork ZODIAC.

Printed and bound in Spain

CONTENTS

Introduction 7

Chapter 1: Aliens

Close Encounters 9 • Vanished Without a Trace 13 • Abducted by Aliens? 17

The Great Alien Cover Up? 21 • Crop Circles 25 • Journeys Through Time 29

Chapter 2: Spirits

Paranormal Projections 33 • Phantom Hitch-hikers 37 • Star Spectres 41

Plagued by Poltergeists 45 • Ghostly Visitations 49 • Haunted Premises 53

Homely Spirits 57 • Ghosts in the Machine 61

Chapter 3: Psychics

Is Anybody There? 65 • Tuned In to the Afterlife? 69

Seeing Into the Past 73 • The Psychic Surgeons 77 • Extra-sensory Perception 81

Divine Inspiration 85 • The Psychic Quest 89 • What Katie Did! 93

Chapter 4: Nature's Mysteries

Mark of the Beast 97 • Mythical Creatures 101

The Mysteries of Shape-shifting 105 • Wisdom from the Deep 109

Nature's Secrets 113 • Steady as a Rock? 117 • Consumed by Fire 121

Chapter 5: The Mind

The Divided Mind 125 • Hidden Powers 129 • Magic Medicine 133

Twin Destinies 137 • Children of Another World? 141

The Exorcist 145 • Torment and Trance 149 • The Miracle Man 153

Chapter 6: Unsolved Mysteries

Lost at Sea 157 • Deadly Contact 161 • Cursed Forever? 165

Paradise Lost 169 • Shipped Through Time and Space? 173

Magical Design 177 • The Search for Gold 181 • An Esoteric Enigma 185

Index 189

INTRODUCTION

It has never, in reality, rained cats and dogs. It has, however, rained goldfish, frogs, small pebbles, asbestos, arrows and snakes! Such showers have been witnessed and recorded for centuries, but were largely ignored by science — until Charles Hoy Fort, a writer in the early 1900s, began to collect those stories that made scientists uncomfortable, those tales that could not be explained or solved by conventional theories. Fort was the pioneer in the studies of paranormal subjects and his books had an indisputable influence on the development of science fiction.

ALIENS UFOs were a recurring theme in Fort's writings, although the term UFO, or unidentified flying object, was not coined until many years after his death. He recorded thousands of cases of strange lights or 'aeroplanes', many travelling at such speeds that they could not have been human in origin. Reportings of UFOs, cases of people vanishing without trace and other instances of possible alien activity on Earth are described in *Chapter 1: Aliens*.

SPIRITS Every year there are reported sightings of apparitions of another kind: ghosts. *Chapter 2: Spirits* shows how they appear in all kinds of places, from the traditional crumbling old house to modern office blocks. Reports of phantoms make eerie reading even though many such ghosts are harmless. Poltergeists, on the other hand, are troublemaking ghosts which hurl objects, start fires or generally terrorise people. Evidence for their existence is not always mere hearsay: spirit-like images have been captured on film.

PSYCHICS Can a communication between two minds be the result of anything other than mere coincidence? *Chapter 3: Psychics* explores this fascinating question. People in touch with spirits from the other world may also apparently see into the past or, more bizarrely, perform complicated tasks, such as surgical operations, for which they have had no training. Paranormal powers can have other practical uses – in the location of archaeological sites, for example – and might also explain the genius of Mozart, or the vision of the poet William Blake.

NATURE'S MYSTERIES Nature's mysteries are not confined to people. Animals past and present, real or mythical can also exhibit unusual behaviour. Even inanimate objects such as stones have been reported as appearing in odd, inexplicable places. *Chapter 4: Nature's Mysteries* covers these as well as the truly baffling phenomenon of spontaneous combustion where people have apparently been burnt – sometimes horrifically – from the inside, their clothes left untouched.

THE MIND The subject of *Chapter 5: The Mind* is the possibility that the mind has hidden powers. ESP, or extra-sensory perception, is often present in children, who may lose it when they get older, though it can also be developed deliberately in adults. Personality is not as straightforward as some would like to believe, with cases of reincarnation and multiple personalities, for example, being well documented. Healing powers, relying on the mind-over-matter approach to medicine common in primitive societies, are enjoying something of a revival today.

UNSOLVED MYSTERIES Some phenomena have never been satisfactorily explained and *Chapter 6: Unsolved Mysteries* describes some of these. Ships and planes have vanished, their passengers and crews lost without trace. Such mysteries are not confined to man-made objects. Ever since the first Egyptian tombs were opened, stories of curses have abounded. Did the magical place called Atlantis ever exist and is it waiting to be discovered? Do ancient, sacred sites such as Stonehenge have supernatural strengths?

> These are just some of the many questions that have never satisfactorily been answered by modern science. Fish still fall, lights continue to appear in the sky – the world is still a magical and mysterious place!

ALIENS

CLOSE ENCOUNTERS

Are we visited by beings from other planets – or is it our imagination?

When the biblical prophet Ezekiel wrote of seeing 'four-faced, four-winged creatures that stepped from a whirlwind', was he actually describing a close encounter? Certainly, the number of similar incidents reported worldwide over the last 100 years suggests that meetings with extra-terrestrial visitors are more than just flights of fancy.

Most alien encounters follow a familiar pattern, with witnesses spotting a UFO and sometimes its inhabitants as well. In the more intriguing cases, people are abducted or invited aboard and subsequently lose all track of time. A few claim to have been left hundreds of miles from where they first saw the craft.

Science can offer no proof in support of these stories, yet many of the witnesses are down to earth farmers or professional people not known for a tendency to fantasize. Do their experiences add up to no more than wishful thinking? Or are there really unknown life-forms beyond the stars?

Judging by the descriptions given, most extra-terrestrial visitors are short, but humanoid in shape. They have grey, translucent or putty-coloured faces, frog or cat-like eyes, pointed ears, small mouths and high foreheads, with claws at the end of their long arms.

Their clothes range from green suits with shiny buttons to black gowns and metallic body-suits with helmets or hoods. A few seem to understand all languages;

THE UNEXPLAINED

Police Constable Alan Godfrey describes his alien captors, England, 1982.

three figures – each of them twice the height of a man – wearing headgear topped by antennae. A beam from the ship seemed to turn the streetlights violet and green, and a strange aroma filled the air.

As the driver approached the craft, he suffered a violent burning sensation. Terrified, he ran to a nearby house where his wounds – later found to resemble radiation burns – were treated. The police were called, but the craft had disappeared; all that remained were giant footprints.

More recently, American Travis Walton recalled his 1975 abduction by an alien craft in Arizona so clearly that artists recreated the scene inside the mother ship (see pages 12 and 15).

Isolated incidents such as these are almost impossible to substantiate, but other close encounters have unwittingly met with a much larger audience. In 1896

others communicate in grunts, pictures, or by telepathy. Some carry luminous wands or swords, and can appear or disappear spontaneously.

Witnesses' reactions to close encounters also bear striking similarities. Many people report being inexplicably exhausted for weeks afterwards, while others suffer recurrent nightmares and severe insomnia.

Famous Sightings

In October 1963 an Argentinian truck driver was halted by a blinding light which abruptly materialized into a metal craft some 35 feet high. Beside it were

UFO entity photographed by Police Chief Jeff Greenhaus in Faulkville, Alabama, USA, in 1973.

AN ALIEN ENCOUNTER

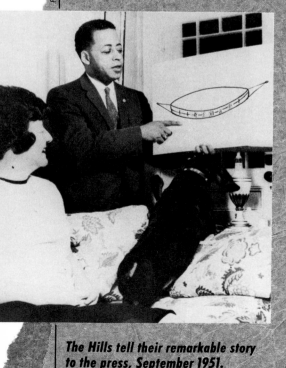

The Hills tell their remarkable story to the press, September 1951.

Many close encounters feature inexplicable time lapses which leave witnesses with only the haziest recollections of what may have taken place.

One of the most extraordinary cases concerns American couple Betty and Barney Hill, who watched a V-shaped craft with red lights at its tips hover in the night sky near Lincoln, New Hampshire in 1951. The craft rose and descended in swoops, making odd buzzing noises. Terrified, they drove away and were later unable to account for a two-hour period.

After dreadful nightmares, the Hills visited a psychiatrist and were placed separately under hypnosis. During these sessions they both revealed that humanoid figures had taken them aboard their spacecraft and kept them there for a considerable time.

Mrs Hill described the aliens as about 1.5 metres (five feet) tall, with long noses, grey complexions, bluish lips and dark hair. Her husband remembered metallic-looking grey skin, nostrils, and eyes which continued a little way around their heads. While Mrs Hill thought the creatures spoke English, her husband said they talked in another tongue which he somehow understood.

Mrs Hill also recalled having a needle inserted into her navel, and went on to explain that the pain apparently disappeared after a gesture from the alien leader. Both claimed to have seen an interstellar map charting the craft's course through the galaxy.

hundreds of people all over California reported seeing a strange craft containing humanoid figures. Between late 1964 and early 1965, a strange cone-shaped craft some 23 metres (75 feet) high and 38 metres (125 feet) wide was spotted hovering over the White House (local radiation levels were later found to be above normal).

Nearly 5,000 people reported seeing a craft which resembled a flying city flying over New York State between 1982 and 1987. And for several months between 1986 and 1987, the American town of Belleville, Wisconsin was plagued by a mysterious red sphere which was seen hovering below the clouds around dusk.

Mementoes of Earth

There are also strange reports of aliens taking samples on their visits, perhaps to analyse life on earth. In July 1965 a French farmer claimed he stumbled on an egg-shaped craft, the two dwarf pilots of which were busily examining his lavender crop. Another farmer reported meeting similar creatures, who said they were from Mars and were keen to learn about fertilizers. He duly left a bag for them, and when he returned next day the bag had gone.

In Jersey City, USA in 1975, ten people claimed they saw humanoids leave a spherical craft and take samples of soil before disappearing into thin air. Mean-

PRESIDENTIAL SIGHTINGS

Former American President Jimmy Carter was quite convinced that he saw what he described as a 'flying saucer'. The sighting took place in the company of several others on January 6th 1969 when he was Governor of Georgia.

It is widely believed that

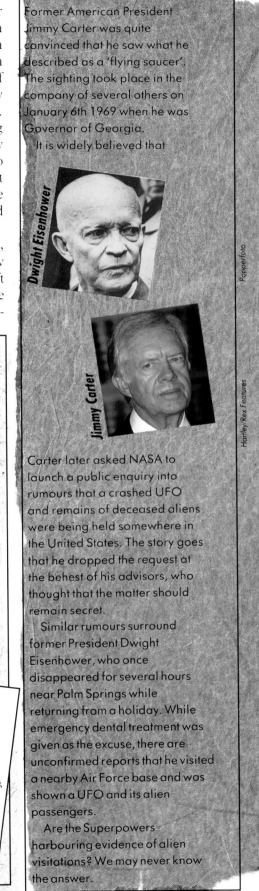

Carter later asked NASA to launch a public enquiry into rumours that a crashed UFO and remains of deceased aliens were being held somewhere in the United States. The story goes that he dropped the request at the behest of his advisors, who thought that the matter should remain secret.

Similar rumours surround former President Dwight Eisenhower, who once disappeared for several hours near Palm Springs while returning from a holiday. While emergency dental treatment was given as the excuse, there are unconfirmed reports that he visited a nearby Air Force base and was shown a UFO and its alien passengers.

Are the Superpowers harbouring evidence of alien visitations? We may never know the answer.

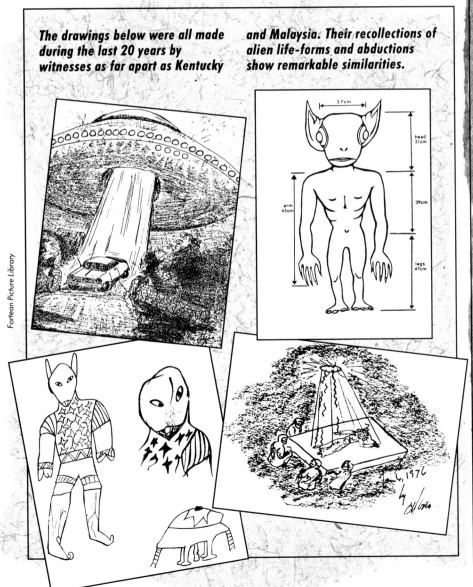

The drawings below were all made during the last 20 years by witnesses as far apart as Kentucky and Malaysia. Their recollections of alien life-forms and abductions show remarkable similarities.

THE UNEXPLAINED

while, in the Midwestern state of Nebraska, a policeman filed a report that humanoids in uniform told him they had been watching the planet for some time and had come to collect electricity!

Assessing the Evidence

The phenomenon of crop circles has given rise to speculation that these huge circular areas of flattened vegetation are in fact the marks left by landing UFOs. Originally thought to be confined to the farmland of southern England, similar circles have since appeared regularly – and inexplicably – all over the world.

Many scientists now believe that the circles are formed by some

Travis Walton with an artist's impression of his visit to an alien spacecraft, Arizona, 1975.

EASTER ISLAND

More than 2,000 miles off the coast of Chile, on the barren hills of Easter Island, stand a group of objects which it is claimed prove that aliens once visited the earth.

The objects in question are giant stone heads, some of which are over nine metres (30 feet) high and weigh more than 50 tonnes. Out of the several hundred scattered stones across the island, those that remain upright stand on equally massive stone burial platforms.

Carved from volcanic rock, the statues date from prehistoric times. Archaeologists believe that they were built by the island's inhabitants in memory of their ancestors. But others point out that without the benefit of a sophisticated pulley system it would have been impossible to haul the statues on to their platforms. Were they perhaps left by beings from another planet whose building skills far exceeded our own?

kind of high energy vortex which is created in the atmosphere under freak weather conditions. Yet this hardly explains the flattened wheatfield discovered by a French farmer in July 1966; not only was the field covered in grease, but a couple reported seeing a brightly lit red sphere hovering over the area the night before.

Yet apart from these isolated phenomena, evidence of UFO landings remains scant. If beings from other worlds are indeed visiting the earth, it seems they are reluctant to leave a calling card.

Photograph of a UFO taken in June 1966 by Paul Villa. According to Villa, the craft measured some 12 metres (40 feet) in diameter.

ALIENS

VANISHED without a trace

Is some strange and mysterious force at work when people disappear from sight – often forever?

The search party was almost ready to give up when, to their complete astonishment, the figure of 14-year-old Edith Horton suddenly staggered out of the Australian bush, covered in bruises and screaming hysterically.

It had been two-and-a-half hours since she had gone missing with three schoolfriends whilst exploring a local beauty spot in the village of Woodend, near Melbourne in Australia. She was suffering from mild concussion and could remember nothing of the past few hours.

When the hunt for the others proved fruitless, the police were called in to assist. But despite extensive searches of the area, the girls were never seen again. They had apparently disappeared into thin air. So began the legend of Picnic at Hanging Rock – one of the world's strangest unsolved mysteries.

St Valentine's Day Mystery

Hanging Rock is an unusual geological formation. Several million years old, it rises abruptly to 152 metres (roughly 500 feet) and pierces the arid skyline in a jumble of miraculously balanced boulders and monoliths from which its name is derived. It was St Valentine's Day, 1900, and Applegate College was enjoying its annual picnic. The party comprised 19 girls, most of them in their teens, and two teachers. One taught French and the other, Greta McCraw, mathematics. They were accompanied by Ben Hussey, a local coachmaster.

At 3pm, three of the senior pupils asked for permission to

13

THE UNEXPLAINED

explore the Rock. The girls – Irma Leopold, Marion Quade and a girl remembered simply as Miranda – were all known to be sensible and responsible. After some discussion between the adults it was agreed to allow them to go, along with Edith Horton, who at the age of 14 was three years their junior.

The girls crossed a nearby stream and disappeared from sight at around 3.30pm. An hour later, Mr Hussey became anxious about gathering up his charges and discovered that Miss McCraw was also missing. As the minutes ticked by with no sign of any of the absentees, irritation quickly turned to consternation and the worried picnickers began a search.

Path to the Rock

They soon found a trail of broken bracken and disturbed scrub. It led to the southern face of the Rock. The search continued for nearly an hour, before the hysterical Edith Horton suddenly emerged from the bush. She could tell her interrogators nothing at all of what had happened to the other three students.

The hunt for the missing girls and their teacher went on for several days. Bloodhounds were given samples of Miss McCraw's clothing and an Aboriginal tracker was enlisted – all to no avail. By the following Thursday, a week after the picnic, the worst was feared.

Miraculous Discovery

But then, to their amazement, the unconscious body of Irma Leopold was discovered on the Rock. Apart from minor cuts and bruises to her head, as well as a few broken fingernails, she seemed to have suffered little as a result of spending over a week in the bush.

Her shoeless feet were clean and unmarked and, despite the fact that her corset was missing, there was no evidence of sexual abuse. Like Edith Horton before her, Irma could not remember anything about her ordeal.

The mystery of what happened at Hanging Rock has given rise to endless speculation. One theory is that the girls were inadvertently involved in some kind of time travel, emerging at another time in the past or future. Others believe that they were sucked into a parallel universe; or that the primeval qualities in the Rock itself spirited the victims away.

The only certainty is that the incident seems set to remain an example of one of countless unsolved disappearances that are reported around the globe every year – many possessing uncanny similarities.

Edith Horton said she had noticed a strange pink cloud at around the time of the disappearances. This is a phenomenon which is often linked to the appearance of UFOs. Could the girls have been abducted by some alien spacecraft?

Hanging Rock: scene of one of the most celebrated disappearances. An Australian film (insets) captured the haunting atmosphere.

ALIENS

Artists' impressions of the fateful night when Travis Walton (inset) was allegedly abducted by odd 'foetus-like' aliens.

Alien Intervention?

It is not so outlandish a concept if one is to believe Travis Walton (see also pages 10 and 12) who claimed to have had contact with alien beings.

In November 1975, he vanished for five days when driving to work with five workmates near Snowflake, Arizona. Seeing a bright light hovering over their truck, Walton felt a sudden inexplicable compulsion to approach the light. Jumping out of the truck, he ran towards it while his companions drove off in terror. When they had calmed down enough to return to the same spot, not a trace could be found of their missing friend. Suspicion for the disappearance fell on the other four. Naturally nobody would believe their story, even though it held up under a lie detector test. But then, like Irma Leopold before him, Walton suddenly reappeared. He was very confused and shaken, but none the worse for his ordeal.

According to Walton, a beam of light had lifted him into a spacecraft, where he was examined by strange 'foetus-like' creatures. Having completed their examinations, they then

---

WORLDS WITHIN WORLDS

Do some people lead an infinite number of lives?

It was the extraordinary accuracy of his dreams which compelled J W Dunne to formulate a series of theories which caused a great deal of excitement in the 1930s. Awakening one morning to find that he had dreamed of a terrible volcanic eruption which had caused the death of 4,000 people, Dunne was astonished to discover on picking up a newspaper that the disaster had actually happened. The only discrepancy was the number of dead, which was ten times his figure.

After some reflection, Dunne decided that he believed people led an infinite number of lives. He became convinced that he was able to tap into his 'parallel existences' when he was asleep. In this way, he was able to delve into the past as well as the future. Dunne's ideas had much in common with Albert Einstein's theories of relativity, which also suggest the existence of strange, parallel worlds.

Could these secret and undiscovered places be the answer to some mysterious disappearances?

THE UNEXPLAINED

returned him unharmed to the place of abduction. It was an alarming experience but Walton can count himself lucky to have returned alive, for history is full of cases of people who have vanished without a trace.

Sudden Disappearances

Almost a century earlier, in the East London area, there was a spate of similar occurrences, which have become known as the 'West Ham disappearances'. One of the first victims was a young girl named Eliza Carter, who vanished mysteriously from her home but later appeared in the street and spoke to some of her friends. They tried to persuade her to go home to her family, but she explained that she could not – 'They' would not let her. She was seen around West Ham for a couple of days before disappearing forever.

Equally extraordinary was the case of Benjamin Bathurst, an employee of the British Foreign Office, who, on 29th November 1809, was about to board a coach outside an inn near Berlin when he vanished completely.

In 1919, the American Charles Hoy Fort included this example in his *Book of the Damned*. Fort also coined the word teleportation – the removal of a person from one place to another by unseen forces. But teleportation does not explain the near-disappearance of US stage magician William Neff. During an appearance at Paramount Theatre, New York, Neff became so translucent that the audience could see the stage curtains through his body. The magician was as puzzled as his viewers. He later admitted to a relative that it was not supposed to be part of his act!

A FAIRYTALE TO REMEMBER

Irish folklore is full of tales of mysterious disappearances. Fairies, elves and demons choose their prey carefully, and once in their grip the victim seldom escapes.

On a tour of Ireland in 1678, a certain Dr Moore fell victim to the irresistible and invisible force of local fairies, who carried him off into the night. The village witch was asked to help and explained that the only way the doctor could escape from their clutches was if he were to abstain from food and drink during his imprisonment. If he ate any of their offerings he would return, but then weaken and die.

The old lady cast her spell and sure enough, as dawn broke, the doctor was a free man again. He explained that he had been offered refreshments during the night but that they had been inexplicably dashed out of his hands. The spell had worked – when morning came, to his relief he discovered that he was alone.

Three witnesses attested to the story, which was signed by one John Cotham. A copy is now preserved in the British Museum.

Beware of fairy-rings and little people, so the story goes, as they are a trap for the unwary!

ABDUCTED BY ALIENS?
CASEBOOK

The creature's deep, dark eyes gave away no clues. Was it a well-meaning friend or a formidable foe?

American thriller writer Whitley Strieber had expected Christmas 1985 to be like any other. He had taken his wife and son to their log cabin in a remote part of New York State, hoping for nothing more than a quiet, cosy celebration with his family.

But on the night of the 26th, Streiber awoke with a start to a strange 'whooshing, swirling' sound downstairs. 'This was no random creak, no settling of the house,' he later remembered. Rather, it sounded more like a large group of people was moving rapidly around.

Within seconds, a fierce little figure peered around the bedroom door. About a metre (three feet) tall, it wore a 'smooth, rounded hat' and a breast plate decorated with concentric circles. Its primitive face had two dark holes for eyes and 'a black down-turning line of a mouth that later became an O.' Strieber watched in horror as it raced towards him. Terrified, he blacked out.

Into the Trees

The author's next recollection was the feeling of floating on air. Although he apparently had no feeling in his body, he realized he was naked and paralysed, with his arms and legs extended as if he had been 'frozen in mid-leap'. In this strange position he felt himself moving down the staircase. He wondered whether he was flying – or being passed along by rows of strange arms.

Strieber then found himself in a woodland clearing. Nearby

THE UNEXPLAINED

stood a creature like the one which had appeared in his bedroom. By Strieber's side was another wearing blue overalls who was doing something to his head. He had the distinct impression that this one was female.

Strieber found himself 'shooting upwards'. Branches whizzed quickly past his face and he noticed the tops of trees rush by. Once again, he blacked out. This time he came to in a 'messy, round room'. Dozens of the tiny creatures were there, moving around in an abrupt, insect-like way that he found repellent.

He noticed that they were about to insert a fine, silver needle into his brain and he began to scream in terror. The female asked him if they could do anything to soothe him. To his utter amazement, he answered: 'You could let me smell you.' The effect was powerful. Once he had inhaled their curious scent – it reminded him of cardboard and sulphur – he realized that he could not be dreaming.

The creatures continued their bizarre physical examination of the author. At one point the female creature inserted an enormous object covered with wires into his rectum. 'I had the impression that I was being raped, and for the first time I felt anger,' he later remembered. Then they made an incision in his forefinger. Strieber lost consciousness for the third and last time.

Trees whizzed past Strieber's face as he shot upwards.

A Tale to Remember

Strange as it sounds, when Whitley Strieber awoke the next morning, he had no recollection of his abduction. Instead, he was filled with a distinct sense of unease and the vivid memory of having seen a barn owl staring at him through the window the night before – a story which he told his family and friends.

Over the next few months he gradually began to feel more and more disturbed. Local gossip

Man meets alien: Strieber's eerie experience captured in a film.

Hollywood's version of the author submitting to medical tests.

about a UFO sighting left him in a state of near panic. Strieber started to remember bits and pieces of his encounter and even began to believe that his wife and child had also been abducted. 'The great and agonizing issue was did it happen or not?' he said later. 'In other words, was I a victim or a madman?'

Did the watchful barn owl mask the memory of the creatures?

Concerned for his sanity, Strieber underwent several sessions of hypnosis where the whole story was revealed. More sinister, he discovered that the Visitors, as he now called them, had been appearing to him since his childhood. Apparently he was – to use modern ufology jargon – 'a recurrent abductee'.

The Truth Revealed?

Strieber was shocked to learn that the Visitors may have affected his family's whole way of life. He wondered whether the many changes of address he had insisted upon were the result of a subconscious desire to escape the creatures. And some of his most vivid recollections – like that of the watchful barn owl – seemed never to have happened. Were they 'screen memories' implanted by the Visitors to mask his encounters with them?

Baffled and disturbed, Strieber wrote *Communion*, his first work of non-fiction, to explain to himself and the world what had happened. Despite his success as the author of *The Wolfen* and *The Hunger*, he found that publishers initially rejected the book with open hostility. However, when it was published in 1987, it rapidly became a bestseller.

Strieber still had to face many critics who accused him of either lying to further his own writing career or succumbing to mental illness. He took a battery of medical and psychological tests. Independent experts said he was not suffering from any kind of illness. And he also passed a lie detector test with flying colours.

Altered States

Whitley Strieber went on to write *Transformation* which was published in 1988. In it, he described

---

A GHOSTLY JOKE?

As Whitley Strieber's many critics have pointed out, he is only too well qualified to 'star' in a psychic thriller about alien abduction. Yet other writers have been placed in similar positions.

One strange tale involves British journalist Frank Smyth. During the 1970s, he made up a ghost tale for a popular weekly magazine. He was amused to receive a post-bag full of letters from people who had supposedly seen the same ghost.

Smyth had written the story to test human gullibility and duly confessed in his publication. So it was with reluctance that some years later he told another ghost story – this time a real one!

Smyth and his friends had been visiting a West Country pub late one night when a girl walked past them. They assumed she had gone to the toilet. Then they realized that there was nothing at all behind them except a blank wall!

Does some cosmic joker set fitting fates for people? Or does an interest in UFO abductions or ghosts act as a magnet in the greater scheme of things?

A ghostly woman at a pub forced the journalist to eat his words.

THE UNEXPLAINED

Whitley Strieber – forced to confront his deepest fears.

Medieval abductions: were fairies simply the Visitors in disguise?

how the Visitors forced him to confront his deepest fears and how he suddenly realized that he both feared and loved them. According to him, further meetings with the Visitors have been mind-altering experiences. He now maintains that they have taught him that the soul can exist without the body.

Searching for Origins

Strieber remains unsure of the Visitors' identity and origins. He concedes that they may come from another planet. But he feels strongly that they most probably come from another level of reality or a different time dimension.

While researching his books, Strieber discovered that bizarre abduction stories were as old as time itself. Many tales of kidnappings by medieval fairies and gnomes seemed to bear an amazing resemblance to his own abduction. However, he believes that the Visitors may actually come from our future – perhaps a group who have learned to travel back in time to study the past. 'They have either been here a long time or they are trying to create that impression,' he says.

Whitley Strieber says his story is not unique. He claims to have met a number of people who have had similar encounters. 'Most remember fierce little figures with eyes that seem to stare into the deepest core of their being. And those eyes are asking for something, perhaps even demanding it,' he remarks.

If this is the case, then only one question remains to be answered – what do the Visitors want? Strieber has a ready, if elusive, answer: 'Whatever it is, it is more than information... It seems to me they seek the very depth of the soul; they seek communion.'

ALL IN THE MIND?

One disturbing aspect of UFO experiences are reports of alien mind manipulation. Many people claim that large tracts of their waking and sleeping lives have been wiped from their memories.

Many contactees learn about their traumatic encounters with aliens during hypnosis. Some even remember levitating and performing other paranormal feats. Such recollections seem so far from the norm that it is tempting to dismiss them as being 'all in the mind'. However, there is often some very real, tantalizing evidence to suggest that the experiences did occur.

Whitley Strieber tries to get to grips with this feeling in *Transformation*. He theorizes: 'There may be quite a real world that exists between thing and thought, moving easily from one to the other – emerging one moment as a physical reality and slipping the next into the shadows.'

UFO contactees tend to dismiss their paranormal experiences.

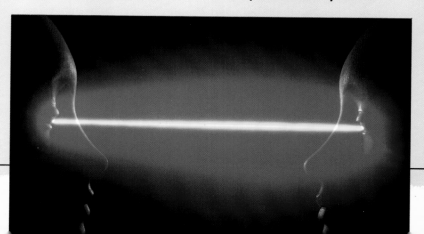

ALIENS

THE GREAT ALIEN COVER UP?

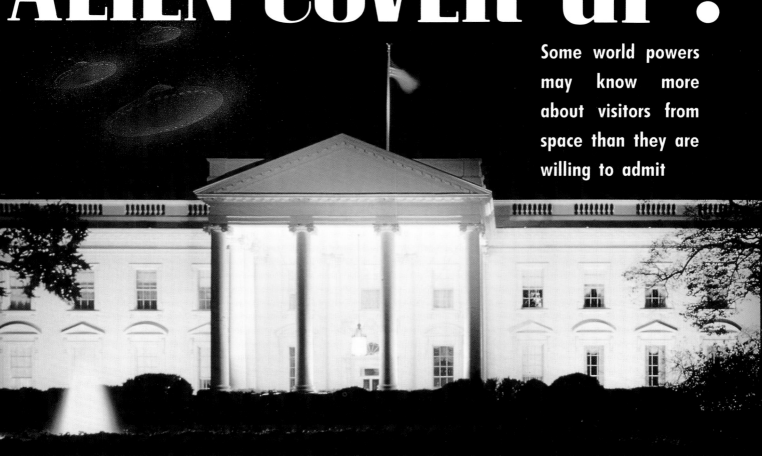

Some world powers may know more about visitors from space than they are willing to admit

The place: Aztec, New Mexico. The year: 1948. The scene: a strange disc-shaped object lands in a lonely field. A team from the United States Air Force (USAF) is sent to investigate.

The 'flying saucer' measures 30 metres (100 feet) across and has no rivets, nuts or welding. Not even a diamond-tipped drill makes a mark on its thin, strong metal casing. Finally, the team notice a broken porthole and spy a knob. Someone pushes it with a pole. A hidden door slides open.

Inside, they find 16 dead humanoids around one metre (three feet) tall, their bodies charred brown by extreme heat. The craft is dismantled and the cabin and bodies are taken all the way to an Air Force base in Dayton, Ohio. There the humanoids are examined and found to be quite similar to human beings – except that they possess perfect teeth!

Versions of the Same?
So went the story told by US newspaper columnist Frank Scully in his 1950 book, *Behind Flying Saucers*. He claimed the Aztec operation was one of four top-secret UFO recoveries. But the USAF immediately denied the story and Scully's theories were ridiculed.

Nearly 40 years passed before another American writer, William Steinman, produced *UFO Crash at Aztec*. He decided that Scully was basically right although some of the details he collected differed.

According to Steinman, the disc was detected by three US radar stations on 25th March, 1948. Its landing point was calculated and a recovery team went to the site.

There they found 14 charred humanoids with over-large heads, oriental features, thin torsos,

THE UNEXPLAINED

slender arms and webbed hands. They had no digestive system and, instead of blood, a colourless liquid which smelled like ozone.

Like Scully, Steinman refused to reveal his sources – laying himself open to charges of fabrication. Then veteran ufologist Leonard Stringfield, a former USAF intelligence officer, dug up more supporting evidence. His sources included an Army intelligence captain who claimed to have cleared a top-secret cable describing the crash and a retired South Florida University professor who had collected testimonies from five people involved in the craft's recovery.

Improbable as the stories sound, nevertheless a common thread emerges – that a UFO landed in New Mexico with some passengers, was recovered and later studied. If this is true, then it seems likely that somewhere documentary evidence exists. Are the facts being kept secret?

Fact or Fantasy?

The theory that some world powers are hiding the truth about UFOs has been circulating since their first highly publicized sightings in 1947. Although most governments have officially denied such stories, there exist many specific examples to the contrary.

For instance, US President Dwight Eisenhower disappeared for several hours while on holiday at Palm Springs, California on 20th February 1954. Some believed he went to Muroc Air Force Base 145 km (90 miles) away to inspect the remains of a UFO. The press were told he had a toothache – but three months later a certain Gerald Light wrote a letter which described his own trip to the base to see five alien aircraft and confirmed the President's visit.

American comedy actor Jackie Gleason is also said to have seen official evidence of UFOs. One night in 1973, he told his wife he had just visited Homestead Air Force Base, Florida, with his friend, President Richard Nixon. There, he saw the remains of aliens that were more than half a metre (between two and three feet) tall with 'large domed heads'. He was told they had been recovered from a craft more than 20 years before.

But perhaps the most telling evidence in support of the view that there is a high-level cover up occurred in 1978. Arizona-based UFO research organization Ground Saucer Watch (GSW) used the Freedom of Information Act to obtain more than 1,000 pages of CIA papers relating to UFO incidents from 1947 to the mid-1970s. This was a substantial collection from an agency which only two years earlier claimed to have had no interest in the subject after 1952!

The Official Line

According to British researcher Timothy Good, there is a long tradition of US investigation into UFOs and related phenomena.

In 1948, a USAF group called

Cover up? Alleged pictures of alien remains at Aztec inspired some sensational speculation.

Since the first sightings, flying saucers have become a mainstay of science fiction films.

ALIENS

Close Encounters: its consultant worked on US Project Bluebook.

Project Sign apparently concluded that UFOs were 'real . . . not visionary or fictitious.' But the report was classified top secret, rejected by the USAF High Command and destroyed.

The aptly named Project Grudge followed, backing journalists who ridiculed UFO sightings until 1950. Then top-secret Project Twinkle was set up to investigate 'intelligent' green fireballs seen in New Mexico, but was closed due to lack of funds.

The 1960s saw the launch of Project Bluebook, headed by intelligence officer Captain Edward Ruppelt. Ruppelt took UFOs

VISIT FROM THE MEN IN BLACK

UFO spotting may be bad for your health. Not only do many contactees develop mysterious illnesses, but some are later visited by sinister Men in Black (MIB) who persuade them to keep quiet.

Although no two stories are exactly the same, many follow a similar pattern. The UFO contactee is visited by a man – or several men – dressed in black suits, crisp white shirts and black ties, often topped off with a 1950s-style black hat. Sometimes they wear dense, black sunglasses that completely obscure their eyes. Most bizarre, however, is the fact that MIB often arrive so quickly after a sighting that it seems unlikely they learned of it through normal means. Occasionally the witness has told no one – yet still the men arrive.

MIB usually turn up in a large black car – a prestigious model such as a Cadillac in the United States, or a Rolls Royce in Britain. These look brand-new, but later checks usually reveal that the number plates are false.

Physically, MIB often look slightly oriental, and sometimes speak with accents to match. Their speech is over-precise and their vocabulary comes straight out of a third-rate gangster film. For instance, one witness was menaced by a MIB who apparently said: 'Again I fear you are not being honest. It would be unwise of you to mail that report.'

That MIB speak and move in an artificial manner has given risen to the theory that they are not the government officials they claim to be, but the aliens themselves!

Alien visitor? A drawing by an eyewitness who was visited by three sinister men in black.

The MOD claim to have no files on British UFOs – such as this flying saucer spotted, drawn and photographed by Stephen Darbishire in February 1954.

THE UNEXPLAINED

seriously, appointing well known scientists – including Dr J. Allen Hynek, who later assisted as a consultant on the Steven Spielberg epic film *Close Encounters of the Third Kind*. But officially Project Blue Book still 'explained away' UFOs.

And the Condon Committee – also in the 1960s – relegated UFOs to the same status as the Loch Ness Monster. Committee head Dr Condon went to his grave maintaining that UFOs were 'bunk'. Nevertheless, before he died he burned all his records.

British Cover Up

According to some researchers such investigations are not just confined to the US. Major studies lasting 30–40 years have definitely been carried out by the governments of Russia and France; Britain, Canada and Australia may also be involved.

British Prime Minister Winston Churchill wrote to his Secretary of State for Air in 1952: 'What does all this stuff about flying saucers amount to? What is the truth?' In reply, he was told that all UFO reports could be explained as illusions, hoaxes, mistaken identification of aircraft, or known weather phenomena.

Even today, the Ministry of Defence (MOD) declares to have no interest in UFOs 'unless they pose a threat to national security – and so far they are not deemed to have done so.' Officially, it does not keep any files on the subject.

With much of the evidence being so specific and yet ultimately so elusive, it is difficult to draw any firm conclusions about what is being covered up. Ufologist Dr

A USAF pilot allegedly took this photo during the Korean war.

Jacques Vallee – an astrophysicist who left his job with the French government in 1988 when he heard of plans to destroy vast numbers of UFO files – puts the case more strongly. 'Something is going on... Something so big that international governments think we would all panic... We have a right to know what it is.'

THE MYSTERY OF RUDLOE MANOR

George Dyer from Birmingham claimed he saw a UFO in 1985. He telephoned London's Ministry of Defence (MOD) to report the sighting and was told to contact a West Country number where the lines were manned 'all the time'.

When British UFO researcher Timothy Good decided to investigate further, MOD officers told him that they were 'unaware of any official research centre'. But a reliable source later told him that the Royal Air Force conducted 'top secret research into UFOs' at Rudloe Manor in Wiltshire. Headquarters of RAF Support Command, some believe it could be a cover for UFO research.

But not all of Rudloe Manor's research is secret. In 1971, for example, a radar expert spent two days tracking an unidentified aerial object. He concluded that it behaved differently from 'any known vehicle or natural phenomenon'.

Ex-Ministry of Defence department head Ralph Noyes had never heard of the establishment. He phoned Rudloe Manor giving details of his Ministry background, adding that he had a UFO report. When he asked if he had contacted the right people, a duty officer replied: 'You've reached the right place'.

However, this does not mean that Rudloe Manor's 30-odd permanent staff are necessarily employed to investigate UFO reports. And it is probably not a good idea to investigate the place in person. In 1985, Wiltshire police arrested Timothy Good for walking around the perimeter of the Manor. He spent several uncomfortable hours in custody trying to explain himself before being released.

Timothy Good found Rudloe Manor hard to penetrate.

Rudloe Manor: research centre?

ALIENS

CROP CIRCLES

Picture from "Circular Evidence" by P. Delgado and C. Andrews/Bloomsbury Press

What is the meaning of the mysterious circles that appear on British farmland? And who – or what – is responsible?

The rolling countryside of southern England sets the scene for a bizarre and thoroughly modern mystery which has become a source of heated debate among scientists, the armed forces, MPs and even ufologists. Each year, from May to September, strange flattened circles appear literally overnight in the crop fields of Hampshire and Wiltshire.

The circles are immediately recognizable by their enormous size and perfect symmetry. They can be photographed, measured, and stringently tested with sophisticated measuring equipment, yet so far no one has found a satisfactory explanation for how or why they exist.

The sudden appearance of crop circles is not the only mysterious thing about them; observers are also baffled by the way in which they appear to be formed.

At first sight, the circles seem

25

THE UNEXPLAINED

to have been cut from the crops with great precision. However, a closer look reveals that the wheat or other plants within them have actually been flattened, with their stems bent downwards at an angle of 90° close to the ground. In this position, the plants concerned continue to grow quite normally until ready for harvesting.

The way in which the plants lie on the circle floor is also peculiar. The flattened stems form clear patterns, swirling outwards from the centre in one direction or the other, or bunching into complex plaited formations.

Pictures from "Circular Evidence"/Bloomsbury

Circular evidence: a famous shot of the Westbury circle, 1987.

Strange Formations

Fascinated by their beauty and mystery, observers began to gather data on the circles. From the mid-1970s onwards, each new discovery was charted and measured, while aerial photographs and soil samples were sent for scientific analysis.

It soon became clear that, far from following a standard pattern, each circle was unique. Sizes ranged in diameter from three to 22 metres (10 to 70 feet), and the number of circles found at any one site varied in every case.

Often, the same field contained a formation of circles – commonly one large central one surrounded by four smaller satellites, or five of similar size grouped together. Other features, such as comet-like 'tails' and additional surrounding rings, were also noted, confounding every theory which the researchers (now called cereologists) put forward.

What is the Answer?

To cynics, the crop circles are no more than an elaborate hoax – the work, perhaps, of high-spirited teenagers. They claim that the circles could be easily recreated using a flail made from a wooden stake and chain, having first tiptoed into the field along the tramways left by crop sprayers.

The cereologists remain unconvinced. Circles have appeared in fields without tramways, and it is virtually impossible to walk through a field full of crops without causing visible damage – especially at night, which is presumably when the hoaxers are supposed to operate.

They also point out that daylight 'hoaxes', staged to ridicule the wilder theories concerning the circles' formation, have resulted

THE DEVIL'S WORK?

Crop circles are thought to be a new phenomenon, being almost unknown before the 1970s. But an illustrated pamphlet of 1678 tells the story of a rich farmer who negotiated with a neighbour over the cost of harvesting a field of oats. The poorer man's price was thought too expensive by the farmer, and he told his neighbour:— the devil himself shall mow the oats before you have anything to do with them'.

The following day, the farmer discovered the crop had been cut. He reported that it was 'as if the devil had a mind to shew dexterity in the art of husbandry ... he cut them in round circles ... with such exactness that it would have taken up above an age for any man to perform what he did in one night.'

Fortrean Picture Library

26

ALIENS

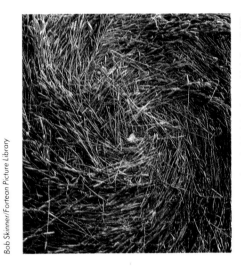

Close up, the flattened stems can be seen to form whorling patterns.

in untidy broken patches which bear little resemblance to the precise shapes of the circles currently under investigation.

One of the most popular scientific theories to account for the mystery has been put forward by meteorologists. They suggest that the circles are created by electrically charged atmospheric vortexes – electric 'whirlwinds' – of a type hitherto unknown and present only in certain conditions. Others claim that since the circles are a relatively recent phenomenon, they must be the result of some sort of military testing. However, there is little other evidence to support this theory, and the armed forces continue to stay silent.

The fact that many of the circles have been found around Avebury, Silbury Hill and Stonehenge – sites associated with ley lines and earth energy points – has prompted some people to propose that they, too, mark areas of abnormally high electromagnetic activity. As yet, though, no one has been able to explain why they occur only in summer, or why some have been found far away from known ley lines.

STANDING STONES

Although few cereologists agree on the origins of crop circles, many believe that unknown energy forces play a large part.

Circles tested with traditional dowsing equipment such as a pendulum have often produced remarkable results. In some cases the apparatus itself was seen to spin in a circle or stiffen like a rod, indicating the presence of considerable energy forces.

Curiously, similar findings have been made at other sites around the British Isles, in particular within the areas bounded by standing stones. At the Merry Maidens near Penzance, dowsers reported the same circling and stiffening of their pendulums. And while some experts believe this was caused by underground springs which cross at certain points beneath the stones, others have put forward more controversial theories.

While the original purpose of standing stone rings has been lost in time, compacted earth around them indicates that dancing or marching may once have occurred there. Because of this, it has been proposed that the stones acted as some kind of accumulator for energy released by the dancers, and that when charged they may have been used to guide landing spacecraft.

Bizarre as this may sound, an experiment carried out on the Rollright Stones in Oxfordshire – the Dragon Project – reported the presence of ultrasonic pulses in the stones at dawn. Stones which did not form part of the circle were tested with no result.

Few scientific conclusions can be drawn from this, but it is clear that some form of energy – whether electrical or magnetic – is present in both standing stones and crop circles. Perhaps science will one day rediscover what our ancestors instinctively knew.

The Rollright Stones, Oxfordshire.

THE UNEXPLAINED

Others, meanwhile, suggest that the circles are marks left after UFO landings, citing other strange events which seem to accompany their appearance.

There are many eye-witness reports of glowing lights, malfunctioning equipment and eerie humming noises from places near the sites of newly formed circles. Strange noises have also appeared on tape recorders without explanation, yet no conclusive links with other extra-terrestrial phenomena have been proved and ufologists themselves are sharply divided on the subject.

Worldwide Discoveries

Since the circles were first reported in the press, others have been found in France, America, Australia, Japan and also other parts of Britain. Each new formation offers cereologists a chance to gather more data, but currently they are no closer to solving the mystery than when their investigations began.

Every year, the approach of summer brings hope that the latest outcrop will leave some startling new evidence or clue as to the circles' origins. As things stand, however, these enigmatic formations are likely to remain a mystery for many years to come.

ELECTROMAGNETIC FIELDS

Thunder and lightning are among nature's most spectacular forces—audible and visual proof of the awesome electrial power present in the atmosphere. These forces influence our moods, behaviour and health; are they responsible for other strange phenomena too?

According to geobiologists, the earth is crossed with a network of electro-magnetic bands which shift and vary in intensity in line with changes in the weather. The crossing points of these bands are known as *geopathic* zones. Healthy plants wither in them, while animals become restless and agitated. Tests have shown that humans, too, suffer a range of ill effects, from general feelings of uneasiness or depression to abnormally high incidences of certain types of cancer.

While there has been much research to find out how electro-magnetism affects living things, scientists still only partly understand the phenomenon. What their work has proved, however, is that 'feeling under the weather' is more than just an old-fashioned expression.

A circle near Cheesefoot, Hants. The lines are made by a tractor.

ALIENS

Journeys through
TIME

Do UFOs challenge the natural physical laws of science? Can our bodies play tricks with time?

One of the earliest written records of the sighting of an unidentified flying object (UFO) is an Egyptian papyrus dating from around 1500 BC. In the intervening 3,500 years, history has documented countless cases of UFOs and other freak phenomena concerning travel in time and space, from time warps and extraterrestrials to fish and frogs falling from the sky.

Freak phenomena are generally an embarrassment to scientists. They prefer to file the paranormal under 'hysteria', 'hallucination', 'misperception', or downright fraud. Yet even an American president has seen a UFO – and confessed to it, as the following example shows.

UFOs

On 6th January 1969, Jimmy Carter and a dozen companions spotted an unidentifiable luminous orb, hovering at 30 degrees above the horizon. 'It was big. It was very bright. It changed

THE UNEXPLAINED

colours and it was about the size of the moon,' said President Carter, a trained nuclear physicist.

Paranormal events challenge our very limited understanding of the concepts of time and space. They violate natural, physical laws – like gravity – and bemuse our logic or expectations. There are things, and people, that go up and do not come down – like St Teresa of Avila, who frequently levitated when she was seized by a spiritual 'rapture'. And there are things that come down which never went up – like showers of herring fry and sardines.

Skyfalls

Such 'skyfalls' have occurred regularly since the era of the biblical plagues. Edible ones are interpreted by some as evidence of interplanetary generosity or a beneficient deity. Sticks and stones and foul-smelling rats, by contrast, are said to be the work of malevolent spirits.

Many individuals claim the ability to transport their consciousness in time and space. Ghosts from the past, prophecies for the future, time warps, dreams, and evidence of past lives gained under hypnosis – all challenge our notion of time as a simple linear dimension, with past, present and future in a sequential order.

Scientists are naturally sceptical about forces which transgress the universal laws. They tend to dismiss 'other-worldly' events as the manifestations of a deluded or hysterical mind. Carl Jung, the psychiatrist, thought

SPACE TRAVELLERS

The Dogon are a West African tribe from the Bandiagara Plateau 480 km (300 miles) south of Timbuktu. Their extraordinary religion centres on the belief that, thousands of years ago, amphibious extra-terrestrial beings called 'Nommos' travelled from the Sirius star system, 8.7 light years away, to educate the tribe.

In the 1930s, two French anthropologists discovered that the 'primitive' Dogon had complex information about the Sirius system which Western astronomers had only recently discovered, using sophisticated telescopes and instruments. They knew, for instance, that Sirius A, the brightest star in the sky, had a very small, white companion star with a 50-year elliptical orbit. The companion star, Sirius B – 'Tolo' to the Dogon – is totally invisible to the naked eye.

In the Dogon faith, Po Tolo was the first star created by God, and is the source of all matter and all souls. They also claim that Sirius A has a second companion star, yet to be discovered by Western astronomers.

that UFOs were a psychological projection of our hopes and fears about the future, for instance.

Hypnosis

Under hypnosis, the stories of people who claim to have had close encounters with aliens, and those of people who are merely told to *imagine* an encounter, are remarkably similar. Many psychologists therefore believe that the mind is somehow 'programmed' to certain stimuli, such as fear and even flying saucers, like the one above, said to have been caught by a camera.

But is it really 'all in the mind'? As yet, conventional science has failed to explain such enigmas as spontaneous appearance and disappearance, prophecy, teleportation, and poltergeists.

Poltergeists

Poltergeists violate the laws of gravity. They levitate beds, hurl furniture, and manoeuvre objects around corners and through solid obstacles (see also page 45). Psychokinesis – the ability to influence inanimate objects by mind-power – is one explanation for poltergeists. Yet people with psychokinetic powers can generally move small objects only a short distance.

The moving of objects or people by unseen forces is called 'teleportation' (see page 16). On 3rd June 1871, the famous Victorian medium, Mrs Guppy, was reputedly teleported to a seance while doing her domestic accounts. Attired in a nightdress, Mrs Guppy landed with a thud on the table and caused quite a disturbance – not surprisingly, since she was known as 'the biggest woman in London' and weighed, it is estimated, more than 230lbs (see also page 66).

Sometimes the teleported person is in two places at the same time. In 1620, Sister Mary Agreda alleged that she made 2,000-mile missionary 'flights' from Spain, to convert the Jumano Indians of Mexico. Her superiors refuted the claims, since she was never physically absent from the convent.

But when the first official Papal missionary visited the

Podbrdo Hill in former Yugoslavia, one of many sites at which a spontaneous visitation of the Virgin Mary has been reported.

Mrs Guppy – Victorian medium, and the 'biggest woman in London'.

THE UNEXPLAINED

Jumano in 1622, he was piqued to find that they had already been instructed in Christianity by a mystery 'lady in blue', and Sister Mary was able to describe the Jumano lifestyle in detail.

The Virgin Mary has made frequent spontaneous visitations, at Lourdes and at Fatima in Portugal. Such manifestations are believed to be evidence of parallel worlds that penetrate our own in space and time. Matter in these 'other-worlds' or spiritual planes vibrates much too fast for us to perceive it, and it is said to be a change in the speed of atomic vibration that can whisk objects and people from one plane to another.

The same vibratory change may account for some of the more mysterious vanishing acts that have occurred in history – though ghosts, gods, demon spirits, fairies and, more recently, UFOs, have all been implicated at some time or other.

Scientists have concluded that no ILE (Intelligent Life Elsewhere) from outside our solar system has any chance of visiting Earth within the next 10,000 years. Others are convinced that high-ranking government personnel are secretly investigating the UFO phenomenon. Whatever the truth, reported sightings are accumulating all over the world, compounding the most intractable mystery of travel in time and space.

ASTRAL TRAVEL

Astral 'projection' happens when the spirit or soul supposedly floats free from the body. 'Out of body' experiences are well-documented and fairly common, particularly among children and adolescents, paraplegics, and victims of accidents or disease who undergo the so-called 'Near Death Experience' when many report viewing their physical bodies from the ceiling of the room as they hover somewhere between life and death.

Knowledge of astral travel seems to be enshrined in the mystical writings of many ancient cultures, including Indian cosmologies, the Jewish Kabbalah, and Egyptian papyri and Books of the Dead, preserved in the tombs of ancient pharaohs and considered a means of gaining access to higher spiritual realms for guidance.

Scientists have often relied on dreams for inspiration, and Edison even slept in his laboratory. Could it be that spectacular 'dream' discoveries, including those of the DNA molecule and the benzene ring, were the result of astral travel?

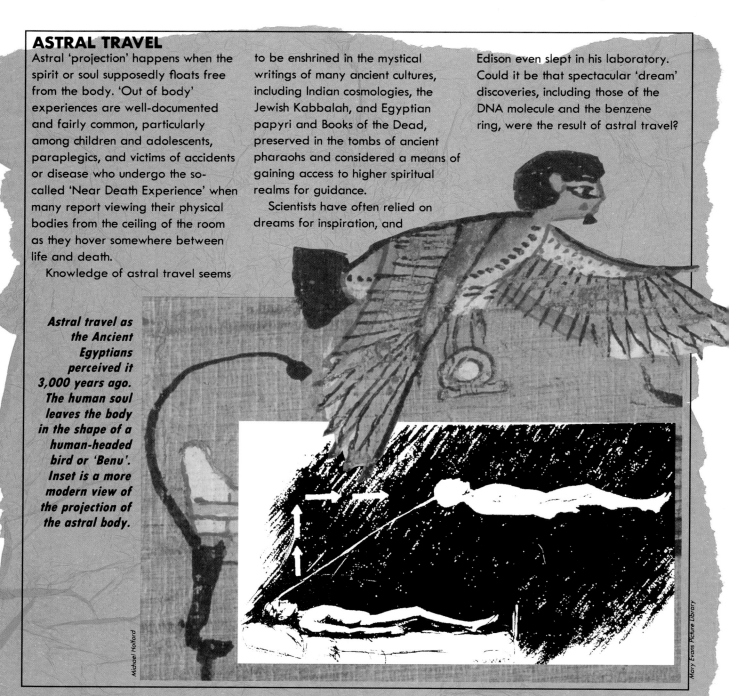

Astral travel as the Ancient Egyptians perceived it 3,000 years ago. The human soul leaves the body in the shape of a human-headed bird or 'Benu'. Inset is a more modern view of the projection of the astral body.

SPIRITS
PARANORMAL PROJECTIONS

Mary Evans Picture Library

Does the camera ever lie – or can spirits and images generated by the mind be captured on photographic film?

Photography may be one of the world's most popular hobbies, but ever since the camera was invented, people have been taking pictures which defy rational explanation. In particular, so-called 'psychic photography' – images of the supernatural captured on film – remains a subject which is surrounded by intense debate.

Although the camera has been known to lie, especially in the hands of tricksters who have something to gain, other cases continue to baffle the experts. Many have been forced to admit that some kind of paranormal phenomenon could be at work.

A celebrated court case of the mid-Victorian era surrounded the work of William H. Mumler, an American photographer who became renowned for his 'spiritual portraits'.

When photographic plates which Mumler had taken of his clients were developed, they often revealed ethereal beings that had not been visible in the studio. After Mumler's work was investigated in 1869 he was charged with fraud, although many people

33

were willing to testify that the blurred additional 'sitters' in the prints resembled relatives or friends who had recently died.

One key witness in the trial was the US Court of Appeals judge John Edmonds, who had been converted to Spiritualism after leading an inquiry into the supernatural. His sworn statement that he had both seen and heard a deceased person in spirit form impressed the jury and helped to secure Mumler's acquittal.

Probing the Paranormal

Mumler's well publicized trial set the stage for further probes into the subject of psychic photography. Although several other well known spirit pictures were proved to be fakes, there were still many which seemed to defy rational explanation.

The respected scientist Alfred Russell Wallace (1823–1913) led the way for research into psychic photography in Great Britain and was convinced that many of the images which he studied were genuine. He noted that the ghostly forms often first appeared during the developing process and believed this was caused by some chemical reaction that the scientific community of the day did not yet understand.

In 1919 the Society for the Study of Supernatural Pictures was established. Most of its members were professional photographers, but among them was the author Sir Arthur Conan Doyle – creator of Sherlock Holmes – who later became involved with many aspects of paranormal and psychic research.

One of Doyle's most controversial cases concerned a famous series of supernatural pictures known as the Cottingley Fairies. These photographs were taken by two Yorkshire schoolchildren called Elsie and Frances, and clearly showed the girls in the company of nymphs and fairies. Doyle was asked to investigate on behalf of the Society, and after a detailed examination of the photographs declared he was satisfied that they were completely genuine.

Unfortunately, the images were later found to be fraudulent. In 1983, Elsie and Frances finally confessed that they were nothing more than a photographic collage which had been set up as an elaborate practical joke.

Mental Impressions

Meanwhile, the case for the existence of genuinely supernatural images was reinforced by the Italian medium Eusapia Palladino, who demonstrated a form of 'psychic print' that could be pro-

A William Mumler print showing the shadowy form of the deceased president Abe Lincoln.

Elsie Wright's gnome fooled even Sir Arthur Conan Doyle.

duced without the aid of photographic plates and chemicals or other mechanical devices.

At the height of her career, around the turn of the century, this extraordinary woman exhibited a number of paranormal talents, including that of levitation. But one of her most celebrated demonstrations was the ability to produce an impression of herself in a slab of putty that had first been sealed in an airtight container to which she had no access.

Thoughts on Film

More recently, investigators have marvelled at the powers of Ted Serios, a hotel porter from the American Midwest. During the 1960s, Serios discovered that he could transfer his thoughts on to film simply by pointing a camera at his face.

Sceptics who suspected that Serios tampered with his film in the darkroom were staggered to learn that he used a Polaroid Land Camera which produced instant snapshots. Some prints showed only his face, contorted in concentration, but his successful 'thoughtographs' (as the pictures came to be known) were little short of amazing. The people portrayed in them were familiar, and landmarks and buildings were easy to recognize.

The scientific community was anxious to discover the precise nature of Serios' talent. The research, however, proved rather more difficult than anticipated.

Eusapia Palladino conducts an experiment in levitation.

WINNING DREAMS

Backing the horses is another popular pastime which takes on a new twist within the world of the paranormal. Many gifted psychics have 'made a killing' at the races following a precognitive dream.

In 1934, Dame Edith Lyttleton, a gifted psychic and former delegate to the League of Nations, began investigating whether people really could profit from their dreams. During a radio broadcast she urged her audience to contact her with information on the subject, and subsequently verified many of the letters from listeners who wrote to say they had used dreams to place winning bets.

Psychics who are able to apply their dreams to betting are usually offered no more than a clue to the winner's identity. A dream about a tropical island, for instance, could point to a horse called Coconut Boy.

Most say that their dreams cannot be produced to order, and that spontaneity is important. They also stress that any bets placed on psychic feelings should be taken at absolutely minimal stakes – there can never be any guarantees of a winning streak.

Psychics have 'made a killing' at the races following a dream about the winner.

THE UNEXPLAINED

THE BELMEZ FACES

As Eusapia Palladino demonstrated, photographic film is not the only medium for paranormal images. Such was the experience of a Spanish housewife who spent years trying to scrub enigmatic pictures from the floors of her home in Belmez.

It began in 1971, when Maria Pereira found a female face on the kitchen hearth. She covered it with concrete, but it reappeared in the same place and was occasionally seen to change its expression. Soon new faces appeared.

Investigators found that the faces showed themselves even when the area was covered by a fixed plastic plate. Tests were run to see if colouring matter was responsible, but the results proved negative.

Maria Pereira had a new kitchen built, but to no avail – the images returned to haunt her there as well. Nobody has ever established the cause of this curious phenomenon, but locals claim that the house was built over a burial site for Christian martyrs killed by Moors during the 11th century.

Serios's thoughtographs were not always successful, and the man himself went on drinking bouts which caused him to leave town without warning. Even during thoughtographic sessions he usually downed a few pints of beer to help him relax, and he would only press the camera button when he felt he was ready.

Even so, researchers had him examined medically and searched his surroundings for any signs of trickery. Nothing conclusive was ever found to suggest fraud.

Mystery Device

Several investigators, including Professor Eisenbud of the University of Colorado, drew attention to Serios' 'gizmo' – a cylinder which he almost always held near the camera when producing thoughtographs. No one ever came up with an entirely satisfactory explanation for its use, but several people drew the obvious conclusion that it had something to do with the images exposed. Eisenbud emphasized, however, that since Serios could produce images of random subjects suggested by anyone in the room, it was unlikely that advance preparations had been made.

Serios's talents have gradually waned over the years, but others have shown similar ability. Uri Geller produced a thoughtographic self-portrait after a challenge – with a cap still covering the camera lens. There is much more to psychic photography, it seems, than meets the eye.

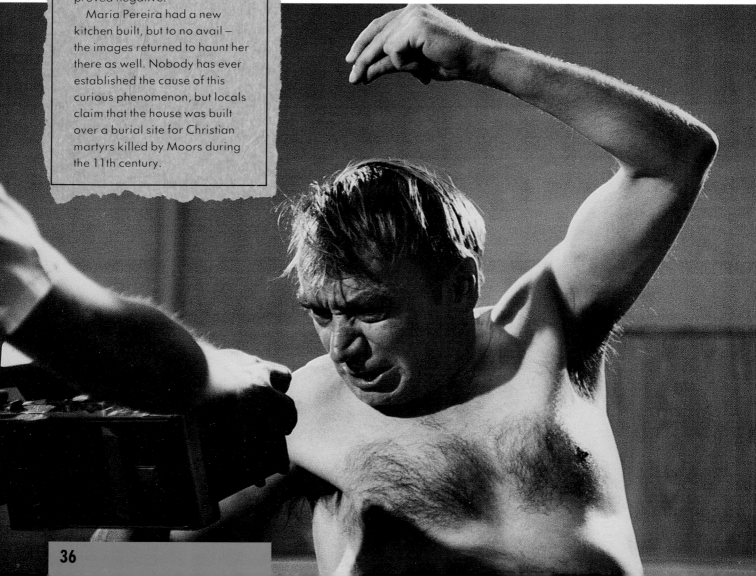

Ted Serios astounded researchers with his 'thoughtographs' – made by pointing a camera at his face.

Gerald Brimacombe/Time-Life Picture Agency/Colorific

SPIRITS

PHANTOM HITCH-HIKERS

Why do some roads seem to be haunted by eerie travellers intent on hitching a lift to nowhere?

Roy Fulton had finished his darts game at a pub in Leighton Buzzard, so he decided to brave the dark, foggy October evening and head for home. The 26-year-old carpet fitter was driving down a desolate road near Stanbridge, Bedfordshire when his headlights caught the form of a pale, young man who was trying to hail a lift.

After the car stopped, the hitch-hiker opened the door and climbed into the back seat without saying a word. He simply pointed in the direction of nearby Dunstable. After a few minutes of silence, Roy decided to break the ice and offer his passenger a cigarette. But when he glanced over his shoulder – the seat was empty. His passenger had simply vanished into thin air!

'I braked, and had a quick look in the back to see if he was there,' Roy said later. 'He wasn't – so I just gripped the steering wheel

37

THE UNEXPLAINED

and drove like hell.' Not even Dunstable police officers could decide what Roy had seen on that night in 1979. But they noted that the driver was sober when he made his report. If he had been hallucinating, then it clearly had nothing to do with alcohol.

Four Varieties of Ghost

The vanishing hitch-hiker is nowadays a much travelled tale – hundreds of similar stories have been told all over the world, earning these phantoms a place in modern folklore. Almost everyone knows a friend (or a friend of a friend!) who cannot resist telling the story of the young couple who meet some nameless horror while driving along a dark, lonely stretch of road...

Nearly 80 cases of US phantom hitch-hikers were analysed by Richard Beardsley and Rosemary Hankey in the California Folklore Quarterly some years ago. According to them, the sightings fell into four basic categories.

Almost half the reports concerned hitch-hikers who gave an address to a driver before disappearing. But on further investigation, the drivers learned that their passengers had died some years previously. In many cases, the very same hitch-hiker had been sighted a number of times by different people.

The next two categories were exclusively female – either young women or old crones. Most of these tales came from male drivers who stopped to pick up a young female traveller. She then proceeded to borrow a coat or a scarf before vanishing. Bizarrely, this clothing was often later found on the woman's grave. Less common were old women who predicted doom and disaster – sometimes even hinting at the end of the world.

The fourth type of story concerned hitch-hikers that were later identified as 'local' gods. For example, the goddess Pelee – guardian of Hawaii's largest volcano Mauna Loa – has appeared frequently in the guise of a hitch-hiker. According to local superstition, it is bad luck to refuse to give her a lift.

Other researchers have noted that ghostly travellers often appear in places where ancient peoples once spotted gods, ghosts or fairies. Swiss psychoanalyst Carl Jung believed that a person's hopes, fears and cultural beliefs determined where the world of myth and the paranormal met. Has the advent of both the car and the hitch-hiker stirred the imagination of modern man and created a new ghost?

Back at the pub, Roy Fulton (third from left) tells friends of his terrifying ordeal on the road to Dunstable.

SPIRITS

Bent on Disaster

Some ghosts, however, go one step further than merely hitching a lift – they throw themselves in front of, or through, moving vehicles. Welsh chef Abderrahman Sennah once had a terrifying experience when he lost his way while driving home from work to Llanidloes, Powys, in July 1973.

As he turned a corner, the head and shoulders of a woman loomed out of the darkness. Horrified, Sennah watched the apparition pass through his windscreen and out through the passenger door. Then he realized he was at The Red Bridge – an old railway crossing just two miles outside Llanidloes – and he drove home shaking in terror. The ghostly woman had apparently been seen many times at the same spot.

The A12 road near Hopton, Norfolk is also haunted by an apparition that is determined to be run over. Andrew Cutajar was driving along it one miserable November night in 1981, when he spotted a grey fog in the middle of the road. On closer inspection, it appeared to be a tall man with grey hair, wearing a long cape and lace-up boots. The man did not move, so Cutajar braked, but then skidded and drove straight through the strange pedestrian. He later described the experience as: 'like going through a cloud.'

The Man in the Macintosh

One of the most bizarre phantom hitch-hiker stories ever recorded in Britain is that of a traveller's eerie repeated meetings with a ghost hell-bent on throwing itself in front of a moving vehicle.

Back in 1958, Exeter lorry driver Harold Unsworth was driving through Wellington, Somerset on the A38 in foul weather. Suddenly, he noticed a man trying to thumb a lift. It was about 3am and the hitch-hiker was soaking wet. Although he wore a macintosh and carried a torch, he had no hat.

The driver felt nothing but sympathy for his fellow traveller and stopped to give him a lift. The man asked to be taken to a destination four miles down the road – and was duly transported there. Unsworth thought nothing more of it, except he noted that the man's conversation was exceptionally gloomy – he talked only of road accidents which had happened recently.

In the course of the next month, Unsworth had two more meetings with the same hitch-hiker. Each time, the man asked to be driven to exactly the same spot and they had the same depressing conversation.

The pair met again a few months later. This time, the hitch-hiker asked Unsworth to wait at the usual place while he

THE LADY VANISHES

Pretty brunette hitch-hiker Maria Roux is most often spotted on the Barandas-Willowmore Road just outside the town of Uniondale, South Africa. But there is just one problem – tragically, she died there in a car accident in April 1968.

South African Army corporal Dawie van Jaarsveld stopped for the woman hitch-hiker while riding his motorcycle one evening in March 1978. Ten miles down the road he felt a bump. Glancing back, he noticed that his pillion passenger had vanished. Terrified that she had fallen off the back of his bike, van Jaarsveld retraced his route. All he found was the helmet he had loaned her, strapped to his luggage rack!

Anton Le Grange also met the hitch-hiker in the same spot two years earlier. The dark-haired girl asked for a lift. Starting his car, Le Grange realized he did not know the street she wanted. He turned to ask her, but she had disappeared.

Le Grange reported the event to the Uniondale police. He and an officer returned to the spot where the girl had disappeared. There, the policeman saw the car door open as though someone was getting out. At the same time, Le Grange heard a chilling scream.

Both Anton Le Grange and Dawie van Jaarsveld identified their passenger as Maria Roux!

Dawie van Jaarsveld on the road again: he swears his passenger was the late Maria Roux (inset).

Inset: Cynthia Hind/Fortean Picture Library

THE UNEXPLAINED

collected some suitcases. When he failed to re-appear after 20 minutes, Unsworth decided to drive on to a transport cafe three miles away.

Harold Unsworth slowed down when he saw a flashing torch light ahead, believing he had spotted a motorist in trouble. Suddenly, he saw the hitch-hiker in the glare of the truck's headlights. Thoroughly frightened, he pulled out to pass the man who, in turn, responded by throwing himself directly in front of the lorry.

Fearing the worst, Unsworth braked hard and then leapt out of his driver's cab. Incredibly, the hitch-hiker was standing in the middle of the road cursing him. As Unsworth watched in utter disbelief, the man simply turned his back and vanished. So did Unsworth. He drove from the scene as fast as he could!

Not wishing to be ridiculed, Harold Unsworth kept his story a secret for 12 years. But when a spate of similar sightings were reported, he finally broke his silence and spoke to a newspaper. But no reasonable explanation ever came to light. This case still remains open to interpretation, as do the other tales of phantoms hitch-hiking to nowhere.

PROPHETS OF DOOM

In one variation of the phantom hitch-hiker story the ghostly traveller brings a message which predicts the end of the world.

One such case followed the eruption of Mount Saint Helens in the American state of Washington in May 1980. Soon, more than 20 people had telephoned the police. Each of them had picked up a female hitch-hiker, dressed either in an expensive white dress or in a nun's habit. She had mumbled about the end of the world and predicted a further eruption in June. This 'Woman of Doom', as she came to be known, could vanish from a car travelling at 60mph. Incredibly, Mount Saint Helens did experience some volcanic activity the very next month.

In the same year, drivers in Oregon met their own prophetic woman, who delivered a similar message before vanishing.

SPIRITS

STAR SPECTRES

Haunting tales of the ghosts who have played an eerie role in the lives of stars of stage and screen

Ghosts are often believed to be an omen of doom, but for the film industry they can be both lucky and lucrative. Write a script that contains a phantom or two, turn it into a film with extravagant special effects, and it is guaranteed to be a success. In recent years, *Ghostbusters* and *Beetlejuice* are just two examples of big box office hits based on this formula. Less well-known, however, are the ghosts that haunt the stars themselves – be they phantoms of the soap opera, spectres of the silver screen or strange apparitions that come and go as and when they please.

Ghost Driver?

American film and television star Telly Savalas, for instance, had one experience that would make anyone's hair stand on end.

Early one morning in the late 1950s, long before he became the lollipop-loving detective, Kojak, Savalas was driving around Long Island, New York. Forced to stop when he ran out of petrol, he decided to walk to the nearest service station for help.

Suddenly a man in a black Cadillac drew up and offered Savalas a lift. When they arrived at the service station, the stranger loaned him money to fill his tank with petrol. The driver even wrote down his name – Harry Agannis – and an address and telephone number when Telly insisted on repaying the loan.

Soon after, the actor phoned the number and Mrs Agannis

41

THE UNEXPLAINED

answered. There was the briefest silence on the line when he asked for Harry. 'He's been dead for three years,' she said. Amazed, Telly decided to visit Mrs Agannis and investigate the mystery a little further.

Savalas learned that his mystery chauffeur's handwriting matched Harry Agannis' script perfectly. The Cadillac's driver had also been wearing a suit identical to the one Harry had been buried in.

Not even Telly Savalas could decide precisely who – or what – he saw that evening. 'I will never forget the incident but I doubt if I will ever be able to explain it,' he said later.

Spectre at Dallas

Equally unforgettable are the hauntings that occur on the film set, especially if they generate masses of publicity. Such was the case with the hugely successful soap opera of the 1980s – Dallas. At the height of the show's popularity, its producers were faced with the task of writing oil baron Jock Ewing out of the script when the veteran actor who played him – Jim Davis – died suddenly. Little did they know that his spirit would soon return to the film studio.

The first to sight Jim Davis' ghost was a photographer who saw him by the swimming pool on the Southfork set, watching the day's filming with obvious enjoyment. Soon after, the apparition developed one of the actor's most familiar habits – straightening Jock Ewing's portrait whenever possible!

Miss Ellie, played by actress Barbara Bel Geddes, was the next to spy the ghost. By chance, she had become ill after the actor's death and was planning to leave the show – when all of a sudden Davis appeared to her.

At first, she thought the apparition of the veteran actor was simply a reflection of Jock Ewing's portrait in the window. She turned and saw nothing, but 'heard' Davis' voice telling her to stick with the show because 'the kids need you'.

Barbara Bel Geddes never doubted that she had been contacted by the spirit of Jim Davis. And she stayed with the show – as he had advised – to the great

SOAP STAR JINXED?

Actress Joan Collins was once reputedly haunted – by a curse!

In *Phantoms of the Soap Opera*, psychic researcher Jenny Randles tells how Joan's troubles began in 1985, after she bought a luxury mansion in Bel Air, Los Angeles. The actress apparently paid a cool three million US dollars for a home that came complete with eight bedrooms, a swimming pool, and a jinx that had supposedly claimed the life of three of its former occupants.

David Janssen, the one-time star of a television series called *The Fugitive* was the last to die at the hands of the curse in 1983. He suffered a massive, fatal heart attack just 48 hours after having a nightmare where he saw his own coffin being carried from the house.

Before him, actor Laurence Harvey – star of the 1950s classic *Room at the Top* – and comedienne Totie Fields had also died there.

The saga continued after Collins moved in. Soon after, she and her husband Peter Holm began divorce proceedings, from which Holm received the house as a part of the settlement.

Was it simply a coincidence that Joan Collins' bad luck began so soon after she and Peter Holm moved into the luxurious house?

Troubled past: Joan Collins' sumptuous Bel Air mansion.

SPIRITS

delight of the many millions of devoted Dallas viewers.

Sceptics might argue that the Dallas haunting was nothing more than a publicity campaign. It certainly earned the show as much attention as the 'Who Shot JR?' debate that kept audiences intrigued during a long break between series.

Fact Or Fantasy?

But are these hauntings just the product of vivid, artistic imaginations? International researcher of the paranormal, Jenny Randles, has collected hundreds of anecdotes from screen and stage stars around the world. She believes that the special sensitivity which helps actors and actresses to 'become' their characters also leaves them open and receptive to other more 'mysterious' happenings.

To make her case, Randles tells the story of British character actress Dorothy Tutin, who had a bizarre experience rehearsing a play called *The Devils*. The story was based on events that happened at Loudun, France in the 17th century, and concerned the exorcism of a group of holy women possessed by demons.

One scene was actually tried out in front of a group of real nuns to see if it rang true before it was performed on stage. 'The extraordinary thing was that the nuns picked it up and we had the experience that happened in the actual play, a sort of catching hysteria,' said Tutin.

Haunted House

Susannah York is another British star who suffered a bad scare. While house hunting, she took her husband to view a spectacular 16th-century home in Essex which came complete with moat and drawbridge.

Both fell in love with it and gained permission from the estate agent to stay the night. But as darkness fell, the actress was overcome by a sense of doom that left her feeling breathless and

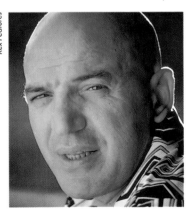

Telly Savalas never forgot his mystery lift in a black Cadillac.

Miss Ellie: convinced Jock Ewing returned to the Southfork set.

Frightened to the point of fainting – actress Susannah York.

trapped. Eventually, she fainted.

Not wanting to give up, the couple returned the next weekend to give the house a second chance – but Susannah experienced the same horrible sensations. This time she also developed an irrational fear that she might be trapped if the drawbridge stuck.

For her own peace of mind, Susannah decided not to buy the house. The estate agents were not surprised. Many people had felt the same way there – apparently because of the ghost of a woman who had drowned after leaping into the moat when the drawbridge stuck fast!

Lucky Ghost

Fortunately, not all tales of star hauntings are full of gloom and doom. Popular film actor Bob Hoskins, who starred in *Mona Lisa* and also in *Who Framed Roger Rabbit*, swears that seeing a ghost changed his life.

The sighting occurred while Hoskins, then an unknown actor, had a job shifting vegetables in London's Covent Garden market. Suddenly, a vision of a medieval nun appeared before him. 'My workmates told me that the ghosts of the nuns are still there in the cellars. If you see one, you are going to have a lucky life,' Hoskins said.

This has come true. Bob Hoskins now has as much work as he wants, a happy marriage and more money in his bank account than he ever thought possible. So who says that seeing a ghost necessarily means that you are in for a spell of bad luck!

The vision of a medieval nun completely changed tough-guy Bob Hoskins' fortunes.

MESSAGES FROM BEYOND

In 1983, the American medium Bill Tenuto claimed that he had spoken to the spirit of the assassinated former Beatle, John Lennon. The story goes that he spoke through a medium during a seance that was recorded for posterity on a tape-recorder which was conveniently to hand.

According to Tenuto, Lennon's spirit was supposedly working with a group called 'the white brotherhood'. This group included the late singer and film star Elvis Presley – and they apparently intended to bring peace and harmony to the world by transmitting song lyrics to anyone who was 'receptive' to them.

As it turned out in the light of recent developments, this peculiar story may have a ring of truth. In 1987, British musical medium Rosemary Brown produced a group of songs, which she claims are new melodies and lyrics from John Lennon. Hitherto, Brown had been best known for receiving classical compositions, which she said were dictated to her at great speed by classical composers such as Beethoven and Liszt.

John Lennon's son, Julian, has also written a song that supposedly stemmed from an after-death pact with his famous father.

SPIRITS

PLAGUED BY POLTERGEISTS

What is the invisible force that sometimes makes chairs move and objects fly through the air?

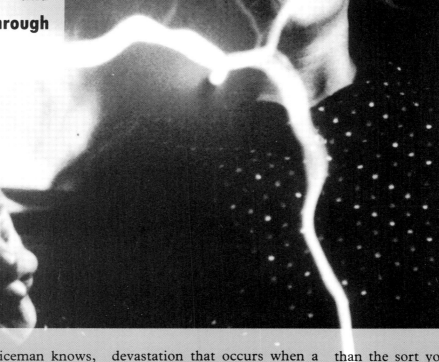

Detail from The Entity/20th Century Fox/Ronald Grant Archive

As any policeman knows, most violent crimes take place within the home, inflicted on one member of a family by another. The stress of being compelled to share a house with one's 'nearest and dearest' – but with whom the only thing you have in common is the tie of family blood – has certainly sometimes ended in murder. Yet there may be other sinister reactions to family tensions – the devastation that occurs when a poltergeist takes up residence, for instance.

Poltergeists differ from ghosts in that they are usually mysterious forces that seem specifically to delight in terrorizing people by hurling objects at them, causing mysterious fires, filling the house with the stench of death, or writing rude graffiti on walls. Poltergeists are invisible vandals and much harder to cope with than the sort you can report to the police, although one family did try to do just that when their furniture moved by itself.

Now regarded as a classic case of destructive haunting, the Enfield poltergeist first made its presence felt in August 1977. A mother was at home with her daughters Janet, aged 11, and Margaret, 13, when a chest-of-drawers suddenly swung away from the wall and a chair moved

45

THE UNEXPLAINED

across the room. Terrified, they brought in a neighbour who called the police. The constable who arrived was later to confirm that something very strange was certainly going on in that otherwise very ordinary London council house.

Acts of Destruction

Over the next 14 months – making this the longest poltergeist attack on record – the family witnessed an enormous range of paranormal phenomena. Fortunately for them, they did not suffer alone, for the case was exhaustively investigated by two leading members of the Society for Psychical Research – author Guy Lyon Playfair, and London businessman Maurice Grosse.

Among the bizarre events recorded was the ripping out of a fireplace by an invisible agency; a curtain that twisted by itself; the hurling of objects – such as the toy brick that hit a photographer in the face when there was no one else in the room – and the teleportation (paranormal dematerialization of an object from one place and its materialization in another) of a book, which went through the wall and ended up next door.

A regular feature of the poltergeist attacks occurred when Janet was pulled out of bed by an invisible force and hurled across the bedroom. Sometimes she ended up asleep on top of the big radio in the far corner, while her sister lay in bed, quivering with fear.

Talking Voices

Janet also seemed to be 'possessed' by a rough, deep male voice, which uttered obscenities. This voice was the last straw for sceptics. The girls were accused of fraud, and Playfair and Grosse were mocked for being easily duped. Yet Guy Playfair had investigated poltergeist activity on many occasions, and both men were there when the Enfield poltergeist was in action, while the sceptics were not.

FAMILY CURSES

There are many examples of family curses, both in fact and in fiction; but the curse that actually befell the Dracula family of what is now Rumania has so captured the imagination that it has been turned into horror fiction.

The 15th-century Transylvanian prince, Vlad Dracula, was a monster in many aspects but real enough. His idea of fun was to impale thousands of prisoners on huge sharpened stakes and then to eat his dinner under their writhing bodies. Even by the brutal standards of the day, Dracula was noted for his savagery. He was cursed many times; and the story goes that his body was discovered to be missing from its grave some years after he was buried. Although there are other explanations, Transylvania is a land of vampire legend, and Dracula was believed to have become one of the 'undead', who crawl from the grave to suck blood from the living.

The legend was revived and embellished in 1897 when Irish theatre manager Bram Stoker wrote a best-selling book entitled simply *Dracula*.

But the story goes on. A distant descendant of the real-life Dracula, the 17th-century Hungarian Countess Elisabeth Bathory, was walled-up alive after it was discovered she had been abducting young girls and bathing in their blood. She believed this kept her young and beautiful. They say her ghost can still be heard screaming.

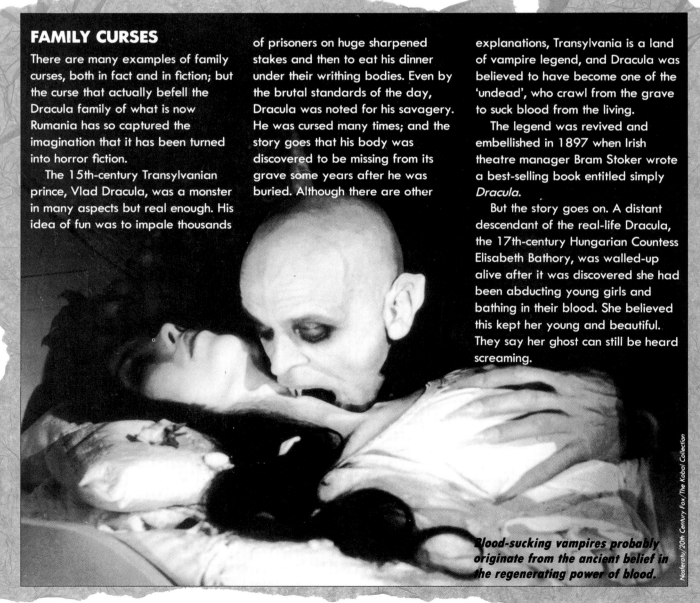

Blood-sucking vampires probably originate from the ancient belief in the regenerating power of blood.

ANCESTOR WORSHIP

To many people, the dead may be gone but they are certainly not forgotten. Indeed, ancestor worship, as practised by certain tribes and religious groups in many forms, ensures that families have a unique bond that straddles generations.

The Ancient Chinese, for instance, believed their ancestors lived on in spirit within their homes; some American Indian tribes held that the dead merely moved a little way from the village; while to the Spiritualists the dead are present all the time, just waiting for a psychic to get in touch with them.

Another example of ancestor-linkage is seen in the 'baptizing of the dead', as practised by the Mormon church. Names of deceased persons are researched, and living Mormons are baptized in the temple on their behalf. Extra sanctity is bestowed on the ceremony by the fact that those being baptized for the dead must first pass a rigorous interview to establish the purity of their lives; and the Mormon temple, unlike their local chapels, is a building that only initiates can enter.

Burning incense, below, is part of Japanese ancestor worship. Part of the Aboriginal ritual to the spirits of their ancestors involves carving tree bark, below right, with mythical animals.

The whole business stopped quite suddenly, just over a year after it had started. In fact, it stopped on the day that Janet had her first menstrual period.

Adolescence

Poltergeist outbreaks, it seems, usually coincide with the physical and emotional upheavals of adolescence, almost as if something is feeding off all the turbulent energy in order to wreak havoc in the home.

Another dramatic case of destructive haunting happened to the Prichard family of Pontefract, Yorkshire, in 1966, when Philip was 15 and his sister Dianne was 12. It started with a mysterious shower of grey dust, halfway between the ceiling and the floor, and a heavy wardrobe that 'walked' across the bedroom floor. After a two-year break, the attacks continued and included mind-boggling phenomena, such as the rearing up like a cobra of a roll of wallpaper. Philip also saw a tall, black-cowled figure during this activity, which is how the case came to be known by the name of The Black Monk of Pontefract.

A priest was called in to exorcize the poltergeist, although he was sceptical about its existence – at least until a candlestick floated in front of his eyes. His visit, however, proved useless. History has proved that poltergeists rarely (if ever) succumb to exorcism.

Mischief-makers

One of the principal features of poltergeists is their sense of mischief, which almost amounts to a capacity for black humour. A Salvationist aunt of the Prichard children tried to drive the force away by singing *Onward Christian Soldiers:* it responded by forming her fur gloves into

THE UNEXPLAINED

a pair of hands and then conducting her as she sang.

But all poltergeist attacks are not as dramatic as these. Sometimes there are simply one or two isolated incidents that recur every time the poltergeist's 'focus' gets upset. Writer-researcher Lynn Picknett has a file of letters describing such 'mini-attacks' sent into her by numerous worried parents and intrigued teenagers.

Interference

Electrical equipment is often the first to show signs of a poltergeist attack, with television interference top of the list. Sometimes it is merely a series of light bulbs that 'blow' during a family row, or water that mysteriously (and repeatedly) floods the kitchen, as if to dampen down the atmosphere.

Objects seem to 'jump' and loud raps, apparently coming from inside the fabric of the furniture, are heard. (Knuckles rapping wood sound very different from a paranormal rap.)

Harmless Spirits

It is all very disturbing and frightening. But the interesting thing is that, with very few exceptions, the poltergeists do no actual harm to people. Objects may be hurled, but only cause a slight bruise. Fires may be started up, but they usually put themselves out as if the poltergeist loses its nerve.

Yet all this bizarre activity does seem to serve a purpose in some respects. For one thing, it is very difficult to carry on a row when pots and pans are hurling themselves past your ear. So do poltergeists cause the type of scene most people would not dare to provoke? Could it be that they sometimes behave like vandals in order to help frustrated individuals get something out of their system?

FAMILY GODS

The Romans, whose gods governed every facet of life, had several deities dedicated to the protection of the household.

The primary function of the two gods known as the Penates (the name derives from the Latin for 'larder') was to preserve the family's food and drink. Indeed, they were so intimately involved in the family's life that, if the family died out, they were believed to disappear, too.

Representations of the Penates were given pride of place in the family hearth, which they shared with other family gods.

The goddess Vesta, who personified domestic fire, presided over the preparation of family meals. Accordingly, she was offered, along with the Penates, the first helping of every meal.

The Lares, who also protected the house, were invoked on all important family occasions such as marriages and funerals; and as she crossed the threshold, a Roman bride always tossed a coin to the Lar assigned to her new household.

The Genius, meanwhile, looked after the marriage bed; appeared at the birth of a baby, being responsible for his or her growth and intellectual awakening; and also formed the newborn's personality.

The Lares were originally gods of agriculture, but became guardians of the hearth.

▲ *Vesta was a popular Roman deity and often appeared on coins.*

SPIRITS

GHOSTLY VISITATIONS

Are ghosts figments of our imagination or spirits trying to communicate?

The spirits of the dead play an important role in all the world's religions and folk traditions, even though religious authorities discourage people from believing that there can be any communication between the inhabitants of this world and the next. For people dissatisfied with the material here-and-now, ghosts form tantalizing links both with the past and with some possible life hereafter. But are these visions just figments of the imagination, or is the dead person really trying to communicate with us?

Hallucinations

Sceptics claim that ghosts are just hallucinations produced in the minds of the percipients (the people who see them). Likely venues for hauntings – ruins steeped in history, old buildings with creaking floor-boards and doors, misty river valleys – work on people's fears and imaginations to create the appropriate phantoms.

The most frequently sighted ghosts are those of well-known men and women who met with a

THE UNEXPLAINED

violent death. Their histories are often familiar to the people who see them, as with the two queens beheaded by Henry VIII: Anne Boleyn, who haunts the Bloody Tower, and Catherine Howard, an even more persistent presence at Hampton Court Palace.

Resident Ghosts

In many other cases, however, the identity of the ghost is not known to its percipient, although the figure seen may subsequently turn out to match the description of some character from local history.

A typical 20th-century ghost was the cyclist seen by a Mr George Dobbs on a cold, snowy night in 1940. To the horror of the unsuspecting Dobbs, who was walking along the road, a car passed straight over the cyclist without its driver seeing him. The cyclist had appeared to Mr Dobbs to have no head, but he had thought this was just a trick of the light. After searching for signs of the accident on the frozen, snow-rutted road, but finding none, he rushed to the pub to tell his tale. There he met the local grave-digger, who was able to inform him that, on that very spot 25 years before, a cyclist had had his head 'torn from his shoulders' in a horrifying motor accident.

The obvious question that springs to mind in cases like this, is why the cyclist's ghost should appear to the unfortunate Dobbs and not to anybody else? Does the ghost have any purpose in appearing to the living or is it just tenaciously fond of anniversaries? In many similar reports

Spiritualists at a seance try to make contact with the dead.

CREATING A GHOST

If ghosts are the product of the mind of the percipient, might it not be possible to create one, as it were, from scratch simply by believing in it very zealously?

This idea occurred to a group of people belonging to the Society for Psychical Research of Toronto, in 1972. First they invented a few dramatic biographical details for their ghost, whom they christened Philip. He was a cavalier, who had killed himself after his gypsy mistress had been denounced by his wife and burnt at the stake as a witch. To give the unhappy Philip a connection with reality, they placed his story in an authentic historical setting, Doddington Manor in Warwickshire, and they hung photographs of the manor around the room where they met to try to meditate Philip into existence.

This proved to be hard work; they had been at it a year before he responded. Even then it was not in the way they had hoped, as a visible materialization. He chose instead to communicate with them by means of rapping and table-turning. The answers he gave were, nevertheless, extremely interesting. He made a number of alterations to the life story they had invented for him.

Making one's own spirit creations is a traditional practice among shamans and magicians in various parts of the world, notably in Tibet, where they are called *tulpas*. The *tulpa* is a thought form, which can normally be sent to do its creator's bidding, but, like Philip, it may start to lead an independent existence.

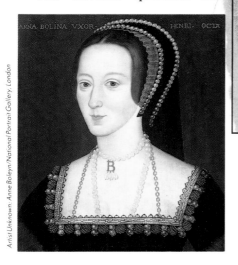

Anne Boleyn, beheaded by her husband King Henry VIII, is said to haunt the Bloody Tower.

the people who see ghosts never discover their identity, despite the most diligent research.

Many modern ghosts do nothing in the least frightening or disturbing. They simply glide noiselessly over the ground or stand at the foot of the bed for a few moments before vanishing. One theory which seeks to explain the rather arbitrary appearances of ghosts is that all past events are somehow recorded in the stones of the landscape or the building where those events took place, and when atmospheric conditions are right or the percipient's mind is tuned to the correct wavelength, then a dim and hazy hologram of history will appear. The more violent the events and the more powerful the emotions involved, the stronger these resonances will be.

Moment-of-death Apparitions

Of greater interest perhaps, and more difficult for sceptics to explain away, are the appearances of people well known to their percipients, especially ones that coincide with the moment of the subject's death. When the Society for Psychical Research was founded in 1882, its principal stated aim in the matter of ghosts was to conduct 'a careful investigation of any reports, resting on strong testimony, regarding apparitions at the moment of death'. The investigation of reputedly haunted houses took second place.

In Victorian England, when so many families were divided, with sons scattered about the far-flung outposts of empire, reports of moment-of-death apparitions were not uncommon. Many followed the lines of the account given by a housekeeper, who in 1856 saw the figure of her son, who was away in India, sit up 'very pale and ashy' from a phantom cart, which she encountered while walking along a country lane. Later she learned that her son had in fact died on the day she had seen his ghost.

Incidents of this kind show that there may well be paranormal means of communication between people, where the bonds of friendship or love are strong and where tragic events are involved. But did the son's ghost appear to the mother to announce his death and bid her farewell, or did the mother somehow sense that her son had died and subconsciously conjure up the vision?

Of course, many worried parents may have similar visions,

▲ *Some ghosts may frighten the living when they appear.*

▼ *Others are a friendly and harmless presence.*

MASS SIGHTINGS OF THE VIRGIN

Although it seems disrespectful to consider sightings of the Virgin Mary in the same way as appearances of secular spirits, the Catholic Church itself is always very careful to investigate thoroughly and sceptically any reported visions of the Mother of God. This is in order to establish that the witnesses have not been deceived by their own imaginations. Only then will the Church give its blessing to the site of an apparition and allow it to become an official place of pilgrimage.

This is especially so when large groups of people claim to have seen the figure of the Virgin. In many cases, like the celebrated series of sightings in the Yugoslavian village of Medjugorje, which began in 1981, she is seen by groups of young children. The initial fear is that the strong personality of one child may work on the others to convince them that they have seen a miraculous vision.

In the case of Medjugorje, the original vision of the Virgin holding the infant Christ first appeared in a meadow to a group of six children, although other people claim to have seen her since. Even more spectacular mass sightings were reported from Egypt between 1968 and 1971, on the roof of a Coptic church in Zeitoun, a surburb of Cairo. Tens of thousands of visitors described a luminous, kneeling figure, her head surrounded by a halo of light.

When one person is granted a vision, it does not really matter whether the image has any physical reality. When many see the same vision, one must consider the possibilities of a miracle, a hoax or powerful forces that link the collective unconscious mind.

In 1961, Loti, a young Spanish girl (second left), had a vision of the Virgin Mary at Grabandal.

which are then forgotten, as their children are not dead at all. Fear and other powerful emotions are an integral part of our relationship with the spirit world. The man racked with guilt, who is haunted by the shape of someone he has killed or wronged, is unlikely to believe that the ghost he sees is 'real'. The extraordinary powers of the human mind offer a 'rational', albeit terrifying, explanation for apparitions like that of Banquo shaking his gory locks at Macbeth in Shakespeare's play.

Harmless Phantoms

Fear and the imagination still play an important part in the materialization of today's ghosts, as in the appearance of phantom hitch-hikers on lonely roads. But the grey ladies, stern-faced cavaliers and hooded monks who haunt the old manor houses, inns and churchyards of Britain usually have nothing at all to say to the living and do not interfere with our lives in any way. They are part of the venerable fixtures and fittings of the places they once inhabited.

SPIRITS

HAUNTED PREMISES

Shops and public houses are often plagued by ghostly apparitions

Like many old buildings, shops are often plagued by ghostly presences. These spirits play tricks by mischievously hiding things, while others make their presence felt by imposing an icy chill on the atmosphere.

One particular Hertfordshire butcher's shop was continually afflicted with mysterious happenings. Fridges repeatedly broke down; fires, lights and radios went on and off of their own accord; and car tyres in the back yard were frequently found punctured, although nobody was ever caught in the act of sabotage.

One day the shop was suddenly filled with acrid smoke, although nothing seemed to be on fire. Eventually, a bag was discovered smouldering away inside a cupboard, although no smoke was to be seen in the immediate vicinity. Even after an investigation into the fire, no trace of combustible materials could be found.

On another occasion, the cloakroom mirror was found shattered on the floor, although

THE UNEXPLAINED

the screws remained firmly in place on the wall. Perhaps strangest of all, a crate of milk bottles would often be found with just one bottle, right in the middle, smashed to pieces. The staff attributed all these events to the mischievous antics of a ghost they nicknamed Charlie.

Visits to the cellar were particularly unpopular. The sudden appearance of a shadowy figure had twice sent young assistants

St Peter's Street in St Albans, Hertfordshire, seems to have had more than its fair share of hauntings and seemingly inexplicable happenings.

tearing up the steps in fright. While passing a pile of cartons in the yard, another youth was startled when a hand appeared over the top and pushed the whole lot on top of him. Needless to say there was nobody to be seen just seconds later.

A female cashier was horrified by heavy breathing near her ear whenever she worked alone in the office. This was occasionally accompanied by the feeling that invisible fingers were gently stroking her face.

The building seems to have had a long history of haunting. For more than 150 years it was the site of The Wellington public house, and in the 17th century it was known as the Blue Boar Inn. It was said that a child living there was accidentally killed by running under the wheels of a coach as it turned into the yard of the inn.

Sense of Humour

A succession of The Wellington's landlords soon found that their pub had its own particular brand of spirit. They all agreed, however, that the ghost seemed to have a playful sense of humour. There would often be urgent rapping on the bar just before opening time, and when the barmaid hurried in she would be puzzled to find the bar completely empty and the outer door still firmly locked.

The beer pumps were sometimes mysteriously turned off and there was often the sound of furniture being dragged around in the living quarters upstairs, although on investigation there was no evidence of anything having been moved.

Lonely Child

A psychical research group held seances at The Wellington in the 1970s, and claimed to have made contact with the spirit of a lonely little boy who told them he used to live at the Blue Boar Inn. He said that the stables in the yard were full of horses and it was there that he played with his friends.

The barmaid of The Wellington seemed to be the prime focus of the ghost's attention. At the seances the boy admitted that he had indeed opened doors for her and liked to stroke her hair and face because she reminded him of his mother.

The psychical research group even suggested that the boy might like to find rest by leaving the premises, but the reply was swift and unmistakable – he was perfectly happy where he was.

Although the pub was sold and refitted as a butcher's shop, and a shopping precinct built over the old stable yard, the ghost of Charlie still seems happy to remain on what he considers to be his home territory.

The Norwich Haunting

A similar mystery surrounds another public house in Norwich, which was also converted into shop premises. The occupants often felt strange cold draughts and caught glimpses of a shadowy figure around the place, accompanied by the sound

A haunted corner of St Albans market place at the turn of the century which is still visited by the ghost of an old lady.

SHOPPING RITUALS

Shop-keepers have always needed luck, and Roman merchants pinned their faith on the benevolent influence of Mercury. The messenger of the gods, Mercury was the Roman deity of commerce, and patron and protector of merchants. On May 15th his festival was celebrated and tradespeople sprinkled themselves with holy water to ensure large profits.

Ritual also plays an important part in modern-day Indian commerce. Choosing the site for a new home or business property must be done on a lucky day and even during a lucky hour. If the proprietors encounter a young virgin along the way this is considered an extra lucky omen.

In China, geomancers use the ancient art of Feng Shui to divine the most favourable sites for their shops. They are concerned that no building should disturb the harmonious flow of the earth's vital energy forces, the Yin and Yang currents, and even in commercialized Taiwan and Hong Kong, Feng Shui can create endless problems when a proposed site falls foul of the laws of this mysterious science.

Shops in the Far East, like most new buildings, have to be sited in a favourable position, according to the principles of Feng Shui.

THE UNEXPLAINED

OLD WIVES' TALES

There are many superstitions and old wives' tales associated with shopping. Market traders often spit for luck on the first coin of the day's takings; an ancient rite to ensure prosperous trade. Shop assistants never date the first sales voucher in advance for fear of jinxing sales.

In the rag trade, superstition forbids twisting wire coat hangers round by the hook, or they may wind up the business. Dusting the counter is equally frowned upon, for fear it may dust away work.

Dust is traditionally associated with luck and money, and some traders believe that if you sweep outwards from the shop front in the morning, you will sweep the day's trade away.

Shoppers should never buy a new broom or even a toothbrush in May, or they may sweep their friends away. When buying a new coat, put money in the right hand pocket when you first wear it, or you may be hard up as long as you have the coat. A new bag or purse should also have a newly minted coin inside for luck, especially if it is a gift, to ensure future wealth.

Friday is the unluckiest day of the week for shopping, especially if it is the 13th. Shoplifters particularly hate Fridays — apparently it is the most likely day to get caught.

Shopkeepers have many superstitions associated with good business. Stall holders spit on the first coin of the day for good luck.

of footsteps on the stairs.

Most of the happenings seemed to centre on an upstairs office, where people often experienced the eerie feeling that they were not alone. A secretary was once alarmed to find her typewriter keys going up and down of their own accord. One explanation was the fact that a woman was said to have been murdered on the premises in Victorian times.

A grocer's shop in Royston, Hertfordshire, had an equally troubled history. Footsteps were often heard in the empty upstairs rooms and on one occasion a shelf full of provisions was found on the floor. Heavy items of stock were often moved around the shop and bottles and jars eerily rattled about for no apparent reason.

When the shop was later taken over by Oxfam, the assistants often found that although they left the shop spic and span at closing time, the next morning clothes would be found in a jumbled mess on the floor.

Most hauntings in shops seem to be related to a death that has happened on the premises when, for some reason, the spirit remains. It is often suggested that this is due to an improper burial or even that it is a punishment for some misdemeanour.

SPIRITS

HOMELY SPIRITS

Why do some houses seem to be singled out for haunting – and what do ghosts hope to achieve by remaining there?

Most people think of a haunted house as a crumbling ruin filled with cold, cavernous rooms and echoing hallways, yet not all ghost sightings occur in such Hollywood-style settings. Ghosts, it seems, can take up residence anywhere, from the humblest cottage to the stateliest of stately homes. No building is immune.

Tricks of the light or mind almost certainly explain many so-called apparitions. It has been shown, for example, that some people experience ghostly hallucinations during the semi-conscious dream-like state that can occur when a person awakens from a deep sleep.

Yet ghost-hunters remain adamant that some houses really are visited by ethereal beings, and the numerous well documented cases of poltergeist activity in existence continue to defy rational explanation. With the number of reported sightings increasing every year, is it possible that so many people could be mistaken?

THE UNEXPLAINED

The famous 'screaming skull' of Bettiscombe Manor in Dorset. Its removal from the house is said to result in blood-curdling screams of anguish.

Stately Hauntings

No self-respecting stately home can afford to be without a ghost, say the sceptics, and in some cases they are probably right. At the same time, it is worth noting that many of these houses have remained in the same family for generations and have particularly strong links with the past.

By common consent the most haunted Royal residence in Great Britain is Glamis Castle in Scotland – family seat of the Earls of Strathmore and once the childhood home of Queen Elizabeth the Queen Mother. Its most famous ghost is a half man-half monster, supposedly born to a former Earl who kept the poor creature hidden from public gaze in a secret room. Despite many attempts, the room has never been found.

Then there is the grey female phantom said to be the ghost of a 17th-century Lady Glamis who allegedly poisoned her husband and was later burned at the stake for witchcraft. The building also boasts its own vampire – an ex-servant who was bricked up in the walls for her misdeeds.

Woburn Abbey, seat of the Dukes of Bedford, is another stately home said to be haunted by a former servant who was maltreated. The spirit makes its presence felt in the television room, where a door opens of its own accord, followed soon after by another across the room, as if the spirit were passing through on some ghostly errand.

The spirits of two of Henry VIII's doomed wives, Jane Seymour and Catherine Howard, are said to stalk the hallways of Hampton Court Palace. His children's nurse, Sybil Penn, has also been seen roaming the building since her grave was moved from nearby Hampton Church, which was destroyed by lightning in 1829. Windsor Castle, meanwhile, is renowned for having a ghost of Henry himself, along with the regal spirits of Charles I (who was beheaded), Elizabeth I and George III.

Among the many haunted political residences across the world is the White House, where former president Abraham Lincoln – himself a psychic – was seen by another former president, Theodore Roosevelt, and by Queen Wilhelmina of the Netherlands.

BORLEY RECTORY

Home to two headless ghosts and a phantom nun, said to have been murdered after an illicit affair, Borley Rectory on the borders of Essex and Suffolk was once reputed to be the most haunted house in Britain. Among the furious poltergeist activity suffered by its owners were whispering and clanging bells, sounds of footsteps, and enigmatic messages scribbled on the walls.

Ghost-hunter Harry Price wrote two books about the strange goings-on at Borley Rectory – reputedly England's most haunted house.

The celebrated ghost-hunter Harry Price was called in to investigate the stories by a national newspaper in 1929, and later produced two best-selling books on the subject. His detective work, though, was criticized by members of the Society for Psychical Research, who found rational explanations for most of the phenomena which Price and others claimed to have witnessed.

Eventually Marianne Foyster, wife of the Reverend Foyster – a one-time occupant – confessed that her husband had dreamed up many of the so-called hauntings in the hope of selling his story to the press. Yet after the building burned down in 1939, reports of ghostly activity around the ruins continued. Could Foyster's stories have perhaps contained a grain of truth?

Yarralumba House in Canberra, home of the Australian Governor General, receives visits from a tiny, dark-skinned ghost, thought to be the spirit of a former Aboriginal servant looking to recover a valuable jewel which legend has it lies hidden there. Even No. 10 Downing Street has a ghost – said to be that of a Regency politician.

Ghostly Missions

Belief in ghosts is often linked to ancient teachings that the soul and the body part company at death, but that the soul may remain 'earthbound' if it is troubled in any way. Certainly, there have been many reports of ghosts which appear to friends and relatives on or around the point of death, as if an attempt was being made by the deceased to set matters straight before departing for the spiritual plane. Other ghosts, it seems, return for revenge on the living.

Such was the likely motive of Christopher Slaughterford, who was hanged in 1709 after a farcical trial for killing his fiancée. Immediately following his death, Slaughterford's ghost appeared at the house of a servant clanking the chains that had bound him in jail. A few days later the servant hanged himself, and it was subsequently found by a court of law that he – not Slaughterford – had murdered the bride-to-be.

Many other houses are said to be haunted by 'screaming skulls' which from time to time emit blood-curdling noises in revenge for not receiving a proper burial. One of the most famous examples resides at Bettiscombe Manor in Dorset, where it has been kept in a bureau for more than 200 years. Removing the skull – which is said to have belonged to a negro slave – from its resting place in the house is claimed to result in blood-curdling screams which send shivers down the spine.

The Berkeley Square Spectre

If ghosts do indeed return to haunt houses in which fearful events have taken place, there must have been some terrible deeds done at 50 Berkeley Square in the heart of London's Mayfair. The ghost in residence at this prestigious address is reputed to have murdered several victims over the years, as well as sending many others insane. One of the most gruesome stories concerns a Sir Robert Warboys who accepted a challenge to spend a night there.

Armed with a gun, plus a bell to sound the alarm, Warboys installed himself in an upstairs bedroom while the owner of the house remained below with a friend. When a shot was heard an hour later the two men dashed upstairs to find Warboys dead, a look of dreadful shock on his face. Since the gunshot had hit a wall, the pair deduced that their friend had died of sheer fright while aiming at whatever it was that had terrified him.

THE UNEXPLAINED

This extraordinary photograph taken at Raynham Hall, Norfolk in 1936 is said by some to be the ghost of Dorothy Walpole – a former lady of the manor who came to a sticky end.

Years later, it seems that the same ghost appeared to two destitute sailors who had taken refuge in the now-empty house. One fled in terror and managed to blurt out his story to a policeman, but when the two men returned to the house they were met by the grisly sight of the second sailor dead, impaled on the wrought iron railings outside.

Among many explanations put forward for the Berkeley Square ghost is that the house was used by forgers, who would go to any lengths to discourage visitors. The theory certainly seems plausible in the light of the case of 'ghostly squatting' concerning a group of monks who were given refuge at a chateau in Chantilly, northern France.

Anxious to stay in their new-found and undeniably luxurious surroundings, the monks hit on the idea of pretending that the place was haunted by disguising themselves as ghosts. The plan was so successful that in due course the chateau was signed over to them – on the condition that they did everything in their religious power to rid the place of the irksome phantoms. Needless to say, the hauntings ceased!

ON THE SCENT OF GHOSTS

Accounts of hauntings generally bear remarkable similarities, with most witnesses reporting hazy, hovering figures, inexplicable knocking noises and sudden chills in the air. Some reports also refer to a pungent odour which fills the room just before the ghost appears, the most famous example being at Hannah House – a 19th-century Indianapolis mansion built by a wealthy American farmer who was vehemently opposed to slavery.

Psychics visiting the house have seen pictures fall from the walls, doors open and shut by themselves, and spoons fly through the air. Noises of clothes rustling and wood crunching have also been heard.

More remarkable, though, is the sickly rotting smell – reminiscent of gangrene – which suddenly arises in certain rooms, causing headaches and nausea. Many attempts have been made to locate the source of the stench, all without success, and cleaning appears to have no effect. Mysteriously, some rooms have also been reported to smell of roses, even though there are none to be found in the immediate vicinity.

The story goes that the mansion was once a link in the 'Underground Railroad' – the escape network by which groups of slaves were smuggled out of the South before the American Civil War. One such group was apparently burned alive in a terrible fire which consumed part of the building. Although no human remains have ever been discovered, this seems hardly surprising in view of the secrecy surrounding the incident.

Hannah House was part of the 'Underground Railroad' network which helped slaves to escape.

SPIRITS

GHOSTS
IN THE MACHINE

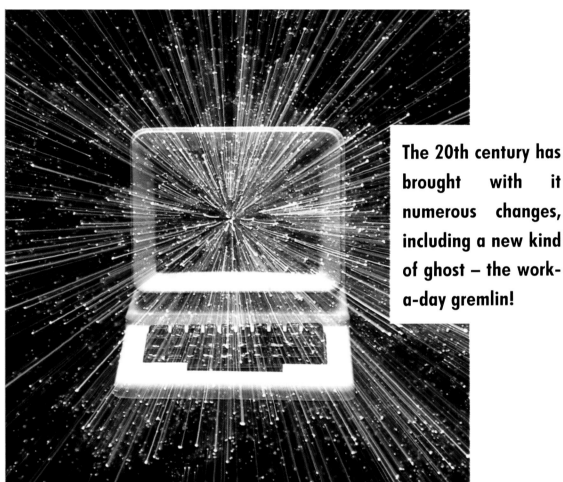

Garry Gray/The Image Bank

The 20th century has brought with it numerous changes, including a new kind of ghost – the work-a-day gremlin!

Most people's idea of a ghost – the ghoulish apparition complete with clanking chains or spectral robes – conjures up images of fog-bound castles criss-crossed with secret passages and burial grounds bathed in eerie moonlight. All far removed, you might think, from the brash, brave new world of the 20th century – but do not be so sure!

It seems that modern technology, public transport and high-rise offices not only provide the wherewithal for thousands of people to earn a living – they also house their own bizarre versions of the traditional ghost-in-residence. And like certain vengeful spirits of the past, some even appear to be life-threatening.

Among many mysterious deaths which have been put down to ghosts in the machine was that of Kenji Urada, a Japanese technician employed by Kawasaki Heavy Industries. In 1981 Urada was repairing a fault in an industrial robot when he was inexplicably beaten to death by the machine's working arm.

Eight years later, a Russian M2-11 supercomputer electrocuted chess Grand Master Nikolai Gudkov just as he was about to checkmate the machine for the third time in a row. Could this be a case of matter over mind?

61

THE UNEXPLAINED

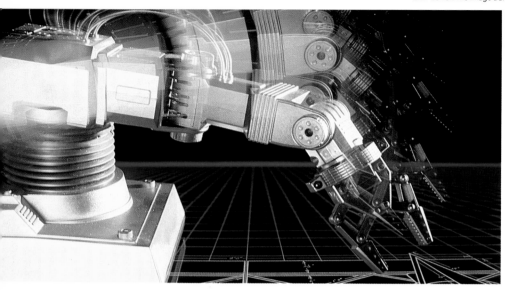

A robotic arm 'murdered' Kenji Urada, a technician, in 1981.

machines like human beings – giving their cars names, cursing the computer that 'refuses' to run, and so on. Most would balk at the idea that we are capable of giving inanimate objects life, yet events recorded by the American biologist Lyall Watson suggest that such occurrences are not as fanciful as they sound.

Psychic Circuitry

In his book, *The Nature of Things*, Watson describes an incident which took place during a filming expedition in South Africa. When recording equipment broke down just before an important interview, Watson's producer waved a worn white overall at it in desperation – only to see the machine whirr back into action. Afterwards it was discovered that the overall belonged to an engineer who could make machines work simply through his physical presence.

Modern science has taught us

All in a Day's Work

A rather more traditional case of haunting in the workplace dates back to 1953, when apprentice plumber Harry Martindale was sent to install central heating in the historic Treasurer's House in York. One day while working in the house he was astounded to see a procession of Roman soldiers walk through a wall and disappear into a pillar on the other

side of the room.

Investigators suggested that the vision was a ghostly remnant of the 'lost' Ninth legion, and interestingly, the Treasurer's House lay on the path of what was once an old Roman road.

Most parapsychologists now subscribe to the view that ghost sightings are projections of past events which have left some kind of psychic imprint on the places in which they occurred. Under the right conditions, so the theory goes, sensitive people unwittingly pick up these psychic vibrations and somehow 'breathe life' back into them, causing a virtual action replay of whatever had occurred there, sometimes long ago.

The theory by no means explains all supernatural phenomena, but it does suggest that man-made objects are capable of absorbing unknown psychic forces. Could the same principle also be applied to so-called 'haunted' computers?

Many people certainly treat

Lyall Watson watched a machine restart after 'recognizing' an engineer's overall.

Harry Martindale's vision of Roman soldiers at the Treasurer's House in York – a more traditional haunting.

SPIRITS

A woman reads a message left by a 16th-century ghost on Ken Webster's 'haunted' computer.

that everything from a stone to a human hand has its own electrical energy field, and that vibrations from one field can affect another either positively or negatively. In the light of Watson's experience, is it too far-fetched to suggest that the technician's overall was still 'charged' with his personal energy, which in the past had been known to have positive effects on faulty machinery?

Computer Talk

Another fascinating case involves a computer installed in an architect's office in Stockport. Video recordings of the antics of the rogue Amstrad PC 1512 were a major attraction at a computer exhibition in London in 1988. Although it was nowhere near a power source, it would switch itself on, begin the start-up procedure, attempt to write on-screen, and then cut out half a minute later with a sound like a groan. One expert believed that the Amstrad was trying to communicate something important.

A more successful instance of computer communication was reported by Ken Webster, an economics teacher from Chester, who received over 300 enigmatic messages on a borrowed BBC microcomputer at his 16th-century home. These were attributed to one Tomas Harden who lived in the house four centuries before. Linguists who have studied the messages have confirmed that the language used matches that of Harden's era.

It seems that we should think twice before commuting to work, taking the lift up to our cheerful, modern office (yes, there are stories of lifts, too, with minds of their own) and switching on the computer. Clearly, nothing in this technological day and age is quite so ordinary and inanimate as we like to assume it to be.

GHOSTS ON THE MOVE

It seems that ghosts on the move are by no means confined to tales of phantom carriages driven by headless horsemen. During the 1930s, for example, a ghostly bus was the talk of London's North Kensington. Locals often saw the vehicle – minus passengers and crew – heading at terrifying speeds towards the junction of St Mark's Road and Cambridge Gardens with its headlights on full beam.

Another intriguing story concerns the famous ghost-hunter Harry Price, who had a strange experience aboard the famous Orient Express. Passing through the German town of Wurzburg, he was rudely awakened at two in the morning by a pistol shot which sounded too close for comfort. Three hours later, he woke again with the sensation that someone was shaking him. There was nobody to be seen.

An attendant told Price that many passengers had spent sleepless nights on the train. He blamed the ghost of a Dutch diamond merchant who in 1923 had decided to steal the packet of diamonds he was supposed to deliver to Budapest. The plan was discovered and the merchant blew his brains out near Wurzburg – in the very compartment of the train occupied by Price.

A phantom bus without passengers was the talk of London's North Kensington during the 1930s.

THE UNEXPLAINED

THE ROSENHEIM CASE

In the summer of 1967, the office of Sigmund Adams, a lawyer in the Bavarian town of Rosenheim, became the focus for some of the most extraordinary paranormal activity ever recorded. It began when Adams received an enormous phone bill, despite the fact that calls were often cut off or interrupted. All four telephones would also ring at the same time but no one answered.

The equipment was replaced but the disturbances continued. A row then broke out between Adams and post office officials when the lawyer learned that the huge bill had been amassed in calls to the talking clock. These had occurred as often as six times a minute – a physical impossibility.

In October, the office lights began to behave strangely; not only did the fluorescent tubes go out with a bang – they also managed to turn 90° in their sockets. Exasperated, Adams had a circuit check carried out by the German Electricity Board on 15th November but no malfunctions were found. The next day, machines to test for voltage fluctuations were installed. Massive power surges which should have blown the fuses were noted – but only during office hours!

Matters worsened. A photocopier leaked, the lights were covered to prevent accidents, and although electricity to the office was now being supplied by a separate generator, the violent phenomena continued. Two investigating physicists admitted that the cause of the problems lay beyond their experience. However, the publicity also caught the attention of Professor Hans Bender of Freiburg's Institute of Paranormal Research.

Searching For Answers

Young Anne-Marie Schneider, who had worked for Sigmund Adams for two years, was as puzzled as her colleagues by the extraordinary events. After noting that the disturbances began when Anne-Marie entered the office at 7.30 am, Bender observed that she was an unhappy, highly strung teenager, frustrated by repressed anger, and with a deep dislike of both her work

Professor Hans Bender established that poltergeist activity was to blame for the Rosenheim phenomena.

and her employer; in short, she was a typical initiator of poltergeist activity.

December and January were difficult months. Paintings turned around on their hooks and pages flew off calendars; drawers slid from desks, office workers received electric shocks and a heavy oak cabinet weighing 180kg (400lb) slid a foot along the floor – without creasing the linoleum.

Anne-Marie left the Konigstrasse 13 office in the middle of January, but the poltergeist activity followed her, and she had to leave several subsequent jobs. In the summer of 1969 her fiancé broke off their engagement when her 'abilities' interfered with the electronic scoring system at his favourite bowling alley.

But the story has a happy ending. Anne-Marie finally moved to Munich, married and started a family. The disturbances ceased.

PSYCHICS

IS ANYBODY THERE?

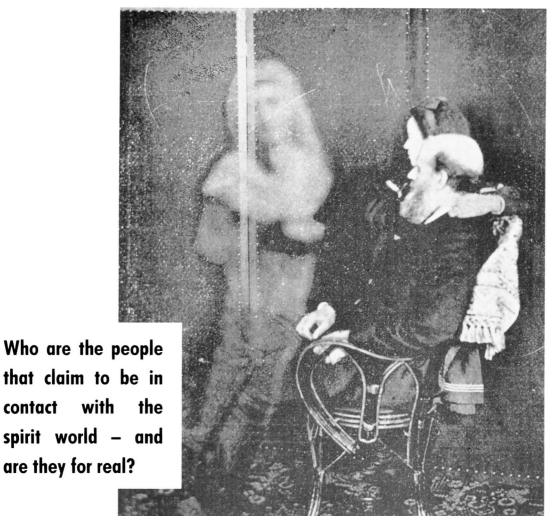

Who are the people that claim to be in contact with the spirit world – and are they for real?

Mankind's fascination with life after death shows no sign of diminishing. Spiritualist churches in Britain and America report growing congregations, and thousands of people worldwide consult those who claim to be in touch with the spirit world.

Researchers at the Institute of Psychiatry in London recently produced evidence to show that many spiritual mediums have suffered damage to the right temporal lobe of the brain. Does this make them sensitive to unknown vibrations which escape the rest of us? Or are sceptics right to call their powers mumbo-jumbo?

Trance Encounters

Central to the medium's work is that mysterious ceremony the seance. when visitors (sitters) gather together to invoke contact with the spirit world. Typically, the medium appears to daydream before entering a state resembling sleep. Then, suddenly, he or she is awake again in the persona of the *control* – the spirit guide with whom the medium has a special rapport.

Messages from the deceased are conveyed via the control, who may cause the medium to speak in a different voice. Some sitters are shocked to see the medium assume the looks, voice or man-

THE UNEXPLAINED

Fraudulent 1930s medium Margery Crandon 'produces' ectoplasm.

nerisms of a deceased friend or relative; others claim to have heard eerie tapping noises or ghostly voices.

Among the bizarre phenomena reported at seances are musical instruments that play themselves, and objects which materialize out of thin air. A few witnesses even claim to have seen *ectoplasm* – a white substance which reputedly oozes from the medium's body.

Spirit Guides

Most people consult mediums after a bereavement, seeking reassurance that a friend or relative is safe on the 'other side'. Others hope to learn a family secret, or to find a missing object whose whereabouts is known only to the deceased.

Such was the case for the family of Edgar Vardy, an inventor who drowned under suspicious circumstances at a swimming pool in Sussex in 1933. Dissatisfied with the coroner's report, one of Vardy's brothers – a member of the Society for Psychical Research – contacted several mediums and attended seances under a false name. While most of the mediums were able to supply remarkably accurate details of Vardy's life – one even named the last book he read – none could throw further light on how or why he drowned.

More successful was the 18th-century Swedish philosopher and psychic Emanuel Swedenborg, who in 1761 located missing papers belonging to the late Dutch ambassador, apparently after conferring with him on the spiritual plane. And in the 19th century,

APPORTS

One phenomenon reported regularly at seances of the past was the appearance of small objects known as *apports*. Thought to be moved by psychic forces, some apports apparently materialized out of nowhere, while others took the form of existing objects which seemed to be transported through solid matter.

According to witnesses, some mediums exercised a remarkable degree of control over apports. The Victorian medium Stainton Moses, a former English clergyman, often moved small items such as pincushions and perfumes from one part of his house to another. Such was the unpredictability of their flight paths, however, that he was once badly injured when a heavy candlestick flew in from an adjoining room!

By contrast, a contemporary of Stainton Moses, Agnes Nichol, who was more famous under her working name Mrs Guppy, could apparently produce apports on demand. She specialized in flowers, which would fall from the air, still moist and with earth at their roots. As described on page 31, she was herself supposedly 'teleported' from her home to a seance four miles away. Mainly because of her weight this became a standing joke among sceptics.

▶ *The apport – once a common 'occurrence' at seances.*
▼ *Mrs Guppy – famous human apport.*

Cases of physical mediumship have declined dramatically this century as methods for detecting fraud have improved. Some cases, however, remain baffling. The Austrian medium Rudi Schneider consistently produced ghostly hands and strange fogs under the most stringent of test conditions.

PSYCHICS

the Englishman Daniel Douglas Home (1833–1886) reputedly conducted more than 150 successful seances for people from all walks of life, including the Russian royal family.

Among the most famous mediums of the 20th century was Gladys Osborne Leonard, who worked with a child spirit control called Feda. In one strange case a Mrs Hugh Talbot was told by Feda that a page in a small dark, leather book on Semitic languages would tell her about her late husband. Baffled, she was persuaded to search for the item; when it was found, the page specified by Feda turned out to be an ancient text about blissful life after death.

Leonora Piper (1857–1950) of Boston, USA used a spirit control named 'Dr Phinuit' to reveal facts about clients which they swore she could not have known. Carlos Mirabelli (1899–1951) was said to materialize people long dead, and in the early 1900s Marthe Beraud – 'Eva C' – could apparently produce ectoplasm from her mouth.

Psychic Fraud

The fantastic claims made on behalf of mediums over the years have inevitably created a backlash of scepticism. Nor has the spiritualist cause been helped by the numerous mediums who have subsequently been exposed as being frauds.

One of the most celebrated cases concerned American sisters Margaret and Kate Fox, who became famous during the 1850s for receiving messages from the dead tapped out in some form of code. When the pair sold their story to a New York newspaper and confessed to having cheated, it sent shock waves across the entire spiritualist world. Margaret, however, later retracted her statement.

Another famous American medium, Margery Crandon, was ruined during the 1930s after a seance 'attended' by her dead brother Walter. Moving lights

Gladys Osborne Leonard worked through the spirit control Feda.

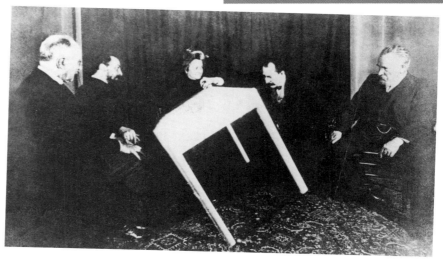

Levitating medium Eusapia Paladino admitted 'occasional' fraud when spirits did not show!

Seances became extremely popular during the 1920s.

THE UNEXPLAINED

Sceptical illusionist Harry Houdini recreates a ghostly hand.

became obsessed with exposing psychic fraudsters after asking the wife of the author Sir Arthur Conan Doyle to contact his dead mother during a seance in Atlantic City. Houdini's suspicions were aroused when a message was received in English – a language his mother had never mastered.

Houdini used his stage shows to demonstrate how fake mediums could reproduce tapping sounds by running a moistened finger across a table, or 'materialize' objects using springs and pulleys. He also showed how so-called 'secret' details of the deceased could be gleaned from obituary notices and tradesmen.

Yet can all reported cases of spiritualist contact be dismissed in this way? Some researchers are convinced that genuine mediums are gifted psychically and somehow extract details from the minds of their sitters. Others have suggested that mediums are schizophrenic, and that their 'controls' are manifestations of a deeply split personality.

Meanwhile, the argument rages on. Before his death in 1979, the American psychic researcher J. Gaither Pratt left a lock with a colleague and promised to try to 'get through from the other side' with the combination, thereby proving contact with a spiritual plane. Several mediums have tried to reach him, but none have been successful.

appeared, and a pair of materialized hands left fresh thumb-prints on a wax pad. Later, the prints were shown to have been made by the medium's dentist – a man who was still very much alive!

During the 1920s and 30s, hundreds of mediums were exposed as charlatans when investigators posing as sitters turned up the lights to reveal stage props, muslin 'ectoplasm' and luminous paint. In 1933 a certain Helen Duncan was fined £10 for stuffing a vest up her dress which was used as ectoplasm; in 1944, she was jailed for a similar offence.

One of the most astonishing confessions of fraud was that of the 19th century Italian medium Eusapia Palladino, who was renowned for levitating tables, elongating her body at will and producing likenesses of the deceased in wet clay. Having been accused of fraud after a series of controlled experiments, she admitted to faking 'occasional' effects – but only when the spirits were reluctant to show!

Harry Houdini, the famous stage magician and escapologist,

PREMONITIONS OF DISASTER

The press coverage following major catastrophes often contains accounts of people who claim to have had premonitions of disaster – whether through precognitive dreaming or simply a hunch.

John Lennon with wife Yoko Ono. The former Beatle's death in New York was predicted over the air.

The 19th-century American president Abraham Lincoln foretold his assassination after a vivid dream and was proved right within only a few days. The sudden death of John Lennon was also predicted when psychic Alex Tanous appeared on an American radio show called 'Unexplained Phenomena' in September 1980. Three months later, the former Beatle fell dead to an assassin's bullet only yards from the studio that broadcast the prediction.

Sometimes those who experience premonitions have time to give a warning, only to have it ignored. In May 1979, after dreaming of a terrible air crash for nine nights in succession, David Booth told the airline concerned of his fears – but to no avail. Ten days after the first nightmare, a DC10 crashed on the runway of the same Chicago airport that had appeared in his recurring dream.

Could it be that time somehow plays tricks on the mind? Or are some of us receptive to warnings from the spiritual plane?

PSYCHICS

TUNED IN TO THE AFTERLIFE?

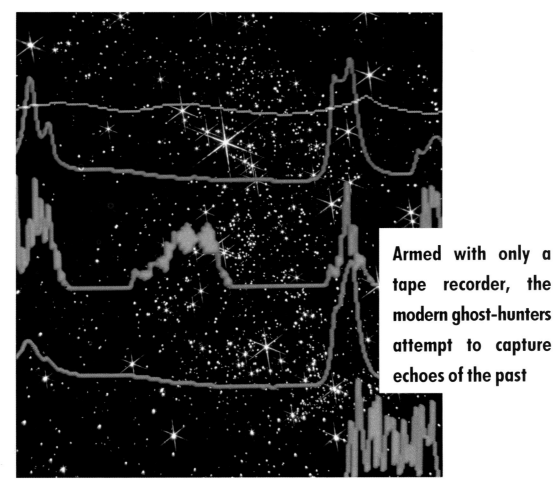

Armed with only a tape recorder, the modern ghost-hunters attempt to capture echoes of the past

Friedrich Jurgenson could hardly believe his ears. The Russian-born film producer, singer and archaeologist had taken advantage of a balmy spring evening in 1959 to record the singing of wild birds from the window of his isolated country home near Stockholm in Sweden. But when he played back the tape, it revealed some surprising anomalies.

'I first heard a finch singing exquisitely,' said Jurgenson, 'then a spluttering, buzzing and voices speaking in Norwegian.' Stranger still, the mysterious conversationalists were actually discussing the nocturnal bird-song!

Ghostly Whispers

But for Jurgenson the clinching experience came a few weeks later when he was playing another tape. A voice which he recognized as his dead mother's suddenly called: 'Friedel, my little Friedel, can you hear me?'

From that moment until his death in 1987, Jurgenson was convinced that the many thousands of voices which he recorded and analysed were direct communications from dead people. There were many who believed him, including the widow of Felix Kersten – the personal doctor of Nazi SS chief Heinrich Himmler. She told one German television documentary that Jurgenson had her dead husband taped, right down to his

THE UNEXPLAINED

'Baltic accent and the whole manner of his voice'.

Jurgenson was not the first person to investigate Electronic Voice Phenomenon – better known as EVP. As early as the 1920s, American inventor Thomas Edison worked on 'a machine or apparatus which could be operated by personalities who have passed on to another sphere of existence'. And famous Italian physicist and inventor Guglielmo Marconi experimented with a similar project until his death in 1937.

However, it was Jurgenson's book – *Voices From Space* – which attracted the attention of a Dr Konstantin Raudive who was living in Uppsala, near Stockholm. So excited was this Latvian-born psychologist and philosopher by the idea of electronic messages that he immediately became an avid recorder.

During his life, Raudive logged over 100,000 voices – among them Swiss psychoanalyst Carl Jung, and other famous names ranging from John F. Kennedy to Russian writer Dostoevsky. He too claimed these were definitely the 'voices of the dead'.

Ghosts in the Machine

One of the perennial questions hanging over EVP is whether the recording equipment is picking up stray voices or static? However, tests have shown that even when the tape recorder is placed in a screened-off enclosure, the voices continue to be heard.

Stickier still is the matter of their interpretation. Not only can it take months for the 'voices' to be recognized, but the listening-in process requires patience and practice. It takes time to accustom the ear to the unusual modulations of EVP. This is made more difficult because most tapes are recorded over a background of 'radio mush' – the noise and static heard between stations. Apparently, this method of recording EVP was recommended by 'Lena', one of Jurgenson's invisible contacts.

The messages are generally short, rapid, rhythmic and cryptic. Large numbers of voices can sometimes be heard in a remarkably short time. Indeed, on one occasion, Raudive claimed to have identified as many as 240 in ten minutes. Some voices sound almost metallic while others speak with great emotion. The quality varies from very poor to exceptionally clear. Most of them, however, are quite weak.

The voices often occur in a variety of different languages – sometimes as many as six in a single sentence. However, those researchers who speak only one language usually find that

Famous words: Dr Raudive contacted (clockwise from top) Kennedy, Dostoevsky and Jung.

communications are limited to their mother tongue. Raudive claimed that this was because the voices use the mind of the investigator as building blocks for communication.

Life After Life?

Others, however, are not so easily convinced. Because the voices are so difficult to hear, critics accuse EVP enthusiasts of self-delusion. Certainly, concentrated listening can lead to auditory hallucinations. And group tests have often resulted in several different interpretations of the same recording!

During his own research into EVP, British parapsychologist David Ellis discovered in the

1970s that many early researchers had not realized that their recorders could pick up snatches of radio broadcasts. He concluded that EVP was nothing more than 'indistinct fragments of radio transmissions, mechanical noises and unnoticed remarks aided by imaginative guesswork and wishful thinking'.

But his opinions are at odds with those of Dr Hans Bender, a professor of parapsychology at Freiburg University who believes to this day that most messages are authentic. He claims that the voices manifest themselves through the investigator who – like a medium – calls them up. Thus, he argues, positive belief creates an auditory 'ouija board'.

Certainly some findings, namely voice print tests on a selection of EVP recordings, have proved that at least some of the voices are genuinely human. Moreover, a number of researchers have noted that some voices actually respond to specific questions or to events that occur at the time. One night, British EVP researcher Gilbert Bonner nodded off after he had set up his recording equipment and slid into a rather inelegant position. When he played back the tape he was astonished to hear a female voice remark: 'Bonner looks quite ridiculous.'

As well as commenting on past events, sometimes the voices can solve unexplained mysteries. One Viennese researcher, Hans Lutsch, apparently spoke to a man called Gunter Barr who was stabbed to death by an unknown assailant. When asked: 'Does your wife know the name of your murderer?' the voice responded: 'It was her.' A few days later, the investigating police officers came to exactly the same conclusion.

Sign of the Times

Although opinions remain divided, the number of EVP investigators continues to grow. There are now literally thousands of

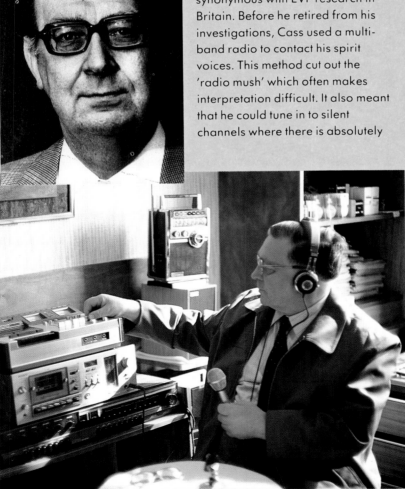

MURMURS FROM BEYOND THE GRAVE

Raymond Cass, a one-time hearing aid salesman from York, is a name synonymous with EVP research in Britain. Before he retired from his investigations, Cass used a multiband radio to contact his spirit voices. This method cut out the 'radio mush' which often makes interpretation difficult. It also meant that he could tune in to silent channels where there is absolutely no chance of picking up fragments of radio transmissions.

Cass claims to have contacted over 5,000 voices. When troubles began in the Middle East in 1973, he received one of his most astonishing messages. Cass had tuned in to his radio equipment when he heard the voice of a girl. 'Today begins the evil struggle,' she said. 'Carefully with nerve gas,' continued the dire prediction of things to come.

On several other occasions, Cass also 'spoke' to the British poet Philip Larkin who died of cancer. Asked what he was doing on the other side, the poet replied 'Just tramping'. This comment convinced Cass that he was talking to his friend who had liked to spend time on his own and often 'tramped' around parks and graveyards to put his thoughts in order.

'Larkin was a man who didn't believe in life after death,' said Cass. 'He would have been very disorientated when he first found himself in limbo.'

British EVP researcher Raymond Cass caught the voice of poet Philip Larkin (inset) on tape.

dedicated amateurs and scientists examining the mystery.

In Germany, some investigators are now working with video cameras trained on fuzzy television screens in an attempt to pick up 'pictures' from the other side. Although some have reported positive results, these have split the ranks and posed yet more questions to be answered.

Perhaps the last word should be left to a former Catholic parapsychologist, Professor Gerbhard Frei, who died in 1967. Dr Raudive tuned in to the voice of the professor one day – when he was feeling slightly doubtful about the validity of EVP. Speaking in a mixture of German and Swedish, Frei told his friend: 'You are sleeping, you do not want to believe. The choice is up to you.'

Talking to the dead – Gilbert Bonner monitors a tape for spirit sounds.

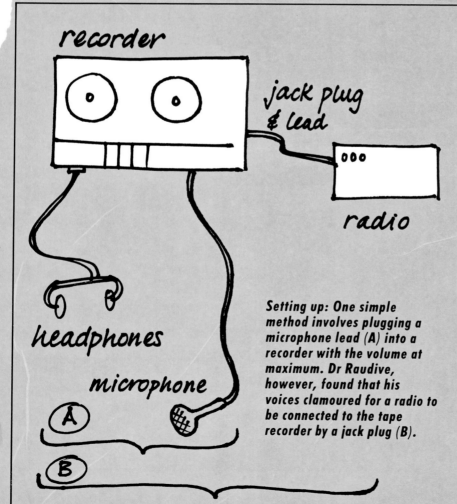

Setting up: One simple method involves plugging a microphone lead (A) into a recorder with the volume at maximum. Dr Raudive, however, found that his voices clamoured for a radio to be connected to the tape recorder by a jack plug (B).

DO-IT-YOURSELF RECORDINGS

Anyone can conduct their own EVP experiments. Only simple equipment is needed – one US investigator taped voices on an inexpensive battery-run recorder!

After setting up your equipment (see left), tune the radio to pick up the 'mush' between stations. Place the microphone as far from the recorder as possible. Plug in the headphones and begin.

Use a brand-new tape, and record for no more than ten minutes. Take a note of the time and date. Make a polite request for the voices to speak. Then remain absolutely silent.

Tape regularly and be patient – it can sometimes take months for voices to come through! If they do, you will need to concentrate because they will be weak. You may need a recorder with a tape counter for the long task of listening and note-taking. Any relevant information should be logged for future reference.

PSYCHICS

SEEING INTO THE PAST

Is it possible that psychics have a part to play in helping many archaeologists unravel the secrets of bygone days?

It was late afternoon on a chill November day in 1907. Behind the closed curtains of a small Bristol architect's office, two middle-aged men were about to conduct an experiment never before attempted in the history of archaeological study.

Frederick Bligh Bond, an expert on religious architecture, was shortly to embark on excavations at Glastonbury Abbey in Somerset. His friend, Captain John Allen Bartlett, also shared his enthusiasm and was to act as a voluntary assistant. The two men's desire to know more about the site – particularly the location of the 16th-century Edgar Chapel, built by the Abbot Richard Beere – was hardly surprising. However, their methods were more than a little out of the ordinary!

The Talking Pencil
Bartlett held a pencil poised over some foolscap paper while Bond placed his hand lightly over his friend's. At the same time the architect spoke sharply into the empty room: 'Can you tell us anything about Glastonbury?'

Almost immediately Bartlett's hand began to move, tracing out a few scrawled lines. The writing was haphazard but not illegible. 'All knowledge is eternal and is available to mental sympathy,' read the message.

With a little encouragement, a 'spirit' which identified itself as 'Brother William' 'drew' a plan of the Abbey, indicating another chapel to the east of the nave.

73

THE UNEXPLAINED

This was described as being nearly 28 metres (65 feet) long and illuminated with azure glass.

Further sittings provided even more details about Glastonbury, and their accuracy astounded the two men once excavations began. The Edgar Chapel, which was uncovered during 1908, was in the exact position and of the precise dimensions predicted. Several pieces of azure glass were also found.

The two men's technique relied on the phenomenon of 'automatic writing' — apparently Bartlett was the channel for spirit guides who took control of his hand. Messages appeared in a mixture of rather strange English and schoolboy Latin. Bond believed that the guides were somehow able to use information already stored within his and Bartlett's own memory banks.

Key to the abbey: spirit writing, allegedly the work of monks who had once lived at the now-ruined Glastonbury Abbey, helped Bligh Bond (below left) piece together the whereabouts of the buried Edgar Chapel (below right).

Tales of Past Lives

Dotted throughout the messages were delightful tales of monastic life provided by the spirit guides, who claimed to be monks resident at Glastonbury during its days of glory. From Abbot Beere to the endearing 'Brother Johannes Bryant' ('the least of all my brethren — save only in my sometime fatness'), they apparently watched over the abbey still, calling themselves alternately the 'Watchers' and 'The Company of Avalon'.

Bond's methods and findings were largely ridiculed at the time, but the archaeologist was to have the last laugh. In a book called *The Gate of Remembrance*, published in 1918, he revealed the location of a new building at Glastonbury — the Loretto Chapel. Excavations over the following two years uncovered fresh ruins near the north transept of the abbey, just as Bond had already described.

The Man with the Pendulum

Bligh Bond's technique is just one which has been used in the name of psychic archaeology. Dowsers, best known for their ability to locate underground water, have also proved useful. Working from maps and in the field, they have achieved incredible results.

One of the most famous was British archaeologist Tom Lethbridge. He believed that past events somehow left psychic imprints on objects and places which any skilled dowser could tune into.

Lethbridge used a pendulum attached to a piece of string which he would dangle over a relic. After years of experimentation in the 1950s, he learned to identify substances by the length of string needed before his pendulum stopped swinging and began to move in circular motions. For silver and 17th-century glazed stoneware (which has a minute silver content) this was 55 centimetres (22 inches); for gold, 74 centimetres (29 inches).

Further researches saw Lethbridge attempt to date old coins according to the number of times his pendulum moved, and even to assess the sex of fossilized sea-urchins! But his claim that Stonehenge had been built in

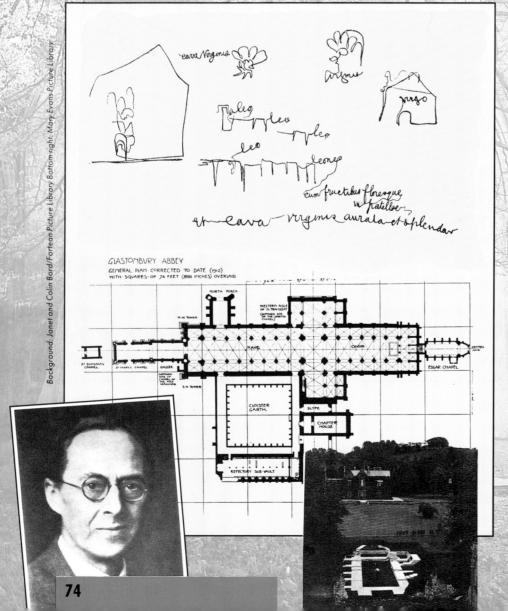

PSYCHICS

2650 B.C. – 650 years earlier than most scientists believed at the time – was later proved correct by radio carbon dating.

During a controlled experiment, Lethbridge also learned that men and women could leave a magnetic impression of their gender on any object – even a stone. He put this theory to the test on slingshots discovered at the Iron Age encampment at Pilsdon Pen, Dorset, and discovered that, with the exception of nine, all had been hurled by males.

Buried Treasures

Other archaeologists look to psychics and psychometrists (object readers) to assist them in the field. Those in Canada turn to truck driver George McMullen – an expert in American Indian sites.

McMullen can give detailed descriptions of treasures buried beneath the ground after just one visit to a site. His mentor, Dr Norman Emerson, president of the Canadian Archaeological Association, claims: 'He quivers and comes alive like a sensitive bird dog scenting his prey'.

Holding a fragment of a 1,000-year-old ring bowl pipe, McMullen was able to say at once who made it and how, along with other details of the Indian lifestyle of the time. McMullen describes himself as 'about 80 per cent

An Italian psychic at work on an archaeological dig earlier this century and (inset) dowser extraordinaire Tom Lethbridge.

Stefan Ossowiecki – was he able to see far into the past?

THE POLISH WIZARD

Stanislaw Poniatowski, Professor of Ethnic Studies at Warsaw University, decided on rather a novel method of archaeological investigation in 1937. In front of six witnesses, he handed 60-year-old psychometrist Stefan Ossowiecki a small flint point.

In the hour that followed, those present heard an uncanny eye-witness description of another age and place: 'Now I see them; the people... rather brownish and quite dark. Hair black. Rather small... Enormous feet. Large hands. Low foreheads... dressed in skins.' Ossowiecki then 'entered' a hut which he said was lit by oil held in a lump of rock. Later he apparently 'saw' a cremation rite which was held for an aged male tribe member.

The psychometrist had been handed a Magdalenian flint dated at 16,000 – 10,000 B.C. His assertion that it came from Belgium or France was correct. In addition, later excavations at Etiolles and Vererie in northern France showed that Stone Age man might well have lived in huts as well as caves, and may even have cremated some of his dead.

In more than 20 sittings held over four years, Ossowiecki gave further descriptions of the life of Stone Age man. Academic archaeologists may consider the results pure fantasy, but not so the families of lost Resistance members who were successfully tracked down by the man known as the 'Polish Wizard'.

THE UNEXPLAINED

accurate', attributing his success to his personal interest in the Iroquois Indian culture.

Strange Impressions

Some psychometric impressions, however, are by no means infallible – as one researcher quickly discovered for himself!

Paranormal investigator and psychologist Lawrence Le Shan was about to send a collection of antiquities to medium Eileen Garrett, founder of the Parapsychology Foundation in New York. At the last moment, he realized he had no box for the final item – a cuneiform tablet said to be of Sumerian origin – and he turned it over to a secretary to pack up.

About two weeks later, Le Shan received a reply from the medium which contained not a shred of archaeological information. Instead, Mrs Garrett provided him with a precise description of the secretary who had packed the tablet – right down to two scars on her body!

With examples like this on record, it is hardly surprising that psychic archaeology has its critics. Marshall McKusick, Professor of Anthropology at the University of Iowa, thinks it is a form of mental aberration 'which offers striking glimpses into the intellectual disorders occurring in our own civilization.'

But many of his colleagues disagree. Like Dr Emerson they believe that they have nothing to lose and everything to gain by using people with paranormal talents. The cost is minimal and the information received often pays off – an added bonus since budgets are usually low.

Of course, the results arrived at by psychic means have yet to be put fully to the scientific test. But perhaps one day, the fantastic statements that are made by psychometrists and dowsers will be proved once and for all by the evidence of the spade.

Red herring? Medium Eileen Garrett divined only the recent history of an ancient tablet!

SECRETS OF THE SPHINX?

Even psychic archaeologists count a 'lunatic fringe' among their number. Many are attracted by the quest for information about the legendary lost land of Atlantis.

Ever since Eric Von Daniken, author of *Chariots of the Gods*, claimed that the limestone formation on the shores of Caribbean island of Bimini was an ancient runway for spacecraft, the island has attracted cranks by the score. Their search continues, despite the fact that the submerged 'cyclopean masonry' has been shown to be of natural origin, a wrecked 'Phoenician galley' bears a New York maker's plate dated 1843, and a so-called 'underwater structure' is in fact a 1930s sponge storage area.

In his *Stones of Atlantis* (1978), David Zink describes a trip he made to Bimini with medium Carol Huffsticker. Not only do they claim to have found the fabled 'Fountain of Youth', but Huffsticker declares she had a clairvoyant vision that the 'causeway' was built by 'star people' from the far-off constellation of the Pleiades 28,000 years ago.

Bizarre as these assertions may sound, they are no less incredible than those made by the highly respected American psychic Edgar Cayce. The 'Sleeping Prophet' believed that the Great Pyramid was built around 10,500 B.C. (2,600 B.C. is the official date). During one reading, he announced that the Hall of Records located in a small tomb lying between the right paw of the Sphinx and the Nile contained artefacts and human remains that would prove Atlantis existed. So far no such vaults have been found...

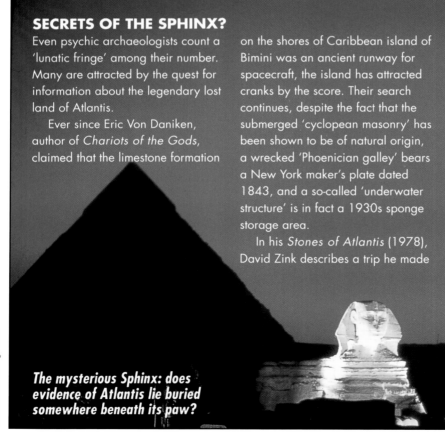

The mysterious Sphinx: does evidence of Atlantis lie buried somewhere beneath its paw?

PSYCHICS

THE PSYCHIC SURGEONS

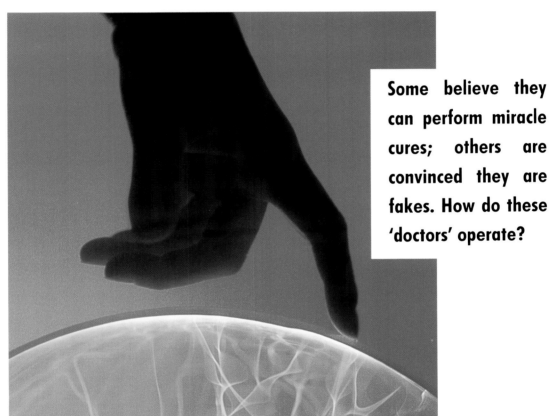

Some believe they can perform miracle cures; others are convinced they are fakes. How do these 'doctors' operate?

Sonja da Costa Cruz was lying flat on the floor while her concerned father and brother-in-law watched over her. She was suffering from an inoperable form of liver cancer. Like thousands of other Brazilians during the late 1950s, she had sought help at a crudely-built clinic in the city of Congonhas, some 250 miles northwest of Rio de Janeiro.

Sonja's life lay in the hands of the barrel-chested, middle-aged 'doctor' with piercing eyes who walked towards her, carrying only a cotton cloth and a rusty tin full of scissors and knives. Despite a complete lack of medical training, Jose Pedro de Freitas – better known as Arigo which means the 'bumpkin' – had already diagnosed Sonja's condition accurately. However, his next actions clearly defied modern medical practices.

Using an ordinary penknife, Arigo sliced cleanly into the woman's flesh. Instead of a massive spurt of blood, only the tiniest trickle appeared. Into the wound, he dropped a pair of scissors which miraculously snipped away all by themselves.

Without hesitation, Arigo plunged his bare hands into the cavity. Moments later, he removed a tumour which he slapped with a dramatic and resounding flourish into the hands of Sonja's flabbergasted brother-in-law! It was as though he was

THE UNEXPLAINED

Jesuit parapsychologist Father Quevedo fakes the skills of the psychic surgeons.

guided by some psychic instinct.

The 'surgeon' continued to astonish his two witnesses who were both medical men. They watched the edges of the wound knit together as Arigo wiped the area with a cloth. At the healer's command, Sonja then stood up feeling weak but suffering no pain. In the months that followed, she gained weight and was finally given a clean bill of health.

Beyond Belief?

As far-fetched as this story sounds, thousands have experienced this psychic surgeon's strange cures first-hand. Not everybody who approached Arigo for an on-the-spot diagnosis succumbed to his crude knife. Some received bizarre but effective prescriptions – combinations of out-of-date medications, or drugs which had not yet arrived on the Brazilian market and vitamins. Arigo's method of prescribing was as instinctive as his operations – he scribbled down prescriptions without hesitation.

American parapsychologist Andrija Puharich investigated Arigo in 1963 and 1968. On one occasion, Arigo successfully removed a tumour from the investigator's arm. Puharich was quickly convinced that he had witnessed real cures – performed with neither anaesthesia nor sterile precautions. He recorded many operations on film.

Puharich also came to believe that during the healings, Arigo was possessed by the ghost of a deceased German doctor called Adolpho Fritz. Apparently, this spirit headed a team of otherworldly medical men. These included a 13th-century monk, Brother Fabianoda Christo, who allegedly created a green ray which somehow removed pain, acted as an instant antiseptic and controlled blood flow.

So strong was each possession, that Arigo was never conscious of performing surgery on any of his patients. On one occasion, he passed out in shock when he later saw a video recording of one of his operations!

Before his death in June 1971, Arigo claimed that he was in fact 'forced' to follow his strange profession. Nightmares, hallucinations, and blinding headaches plagued him until he reluctantly dedicated himself completely to his healing mission.

This psychic surgeon also worked within a strict set of guidelines. Arigo was convinced that accepting any payment for healing would result in the loss of his abilities, and refused to accept even a cup of coffee from a grateful patient. Moreover, he kept up a full-time office job while treating more than 300 people each day.

Age-old Link

Tales like these show the link between psychic surgeons and tribal shamans, whose knowledge and traditions reach back thousands of years. Coincidently, a shamanic vocation often begins

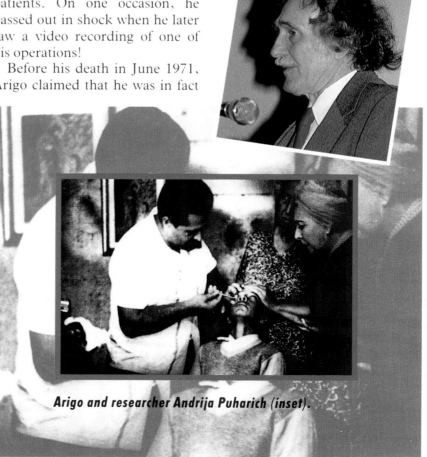

Arigo and researcher Andrija Puharich (inset).

PSYCHICS

Primitive roots: cures often mimic the art of tribal shamans.

the first half of this century, supposedly had two spirit helpers. 'When I treat people those two go right inside a man,' he told anthropologist W. Lloyd Warner. But Willidjungo also believed the local superstition that if he fell into salt water, he would lose his spirit allies.

Keeping the Faith

Psychic surgery today flourishes mainly in Brazil and the Philippines (see also page 151). Researchers attribute this to the strong Spiritist beliefs which still exist in both countries.

These are based on the writings 19th-century French mystic Leon Denizath Hippolyte Rivail who wrote under the pen-name of Allan Kardec. The Spiritist Society acknowledges the presence of another invisible dimension which is populated by souls of the dead, saints and guardian angels. So strong is the faith of its members that 15 years ago, a healing centre in down-town Sao Paulo staffed by 35 voluntary healers saw 482,000 patients annually. A crucial aspect of Spiritist healing calls on patients to recognize that negative thoughts or emotions affect health. Since the subtle interrelationship between mind, body, emotions and spirit is emphasized, many patients leave with a greater sense of personal responsibility as well as feeling physically improved.

Miracle Cures?

Inevitably, critics point to the sleight-of-hand employed by some healers in the Philippines and denounce the whole procedure as a charade. However, the incredible results achieved by

with an illness which immediately stops once the individual accepts his destiny. Shamans also operate within their own set of taboos.

For instance, Willidjungo, a shamanic healer who worked in Australia's Arnhem Land during

DOUBLE THE SPIRIT

'When I was alive, I lived in the east end of London... I was a general surgeon... In the spirit world I wanted to come back... I use simple explanations when I talk with patients. I break them down into (1) the physical and (2) the spiritual which contains energies and gives life to the physical body. I am able to operate on it (the spirit).'

This remarkable statement was supposedly made by a deceased Scots surgeon, Dr William Lang (1852–1937), through his medium George Chapman in December 1969. Lang's family were so convinced by the ghost's identity that they have remained in close contact with Chapman, once a fireman and now a full-time healer.

Watching a Chapman-Lang operation is a strange experience. As 'Dr Lang' operates on the patient's spirit body – some inches clear of the physical one – his medium appears to make hand movements with invisible surgical instruments. Patients see fluids appear out of thin air and hear snapping sounds as if the doctor's team of spirit assistants are passing him instruments. Scars sometimes appear on the patient's body but heal extremely quickly.

Many people have attested that Lang's spirit actually exists. In 1963, Ethel Bailey was cured of a lazy eye by the spirit of Dr Lang – the surgeon she had visited 26 years earlier for the same problem. The pair actually 'recognized' each other!

A man possessed? George Chapman 'operates' on a patient's spirit body in the guise of his alter-ego Dr Lang (inset).

THE UNEXPLAINED

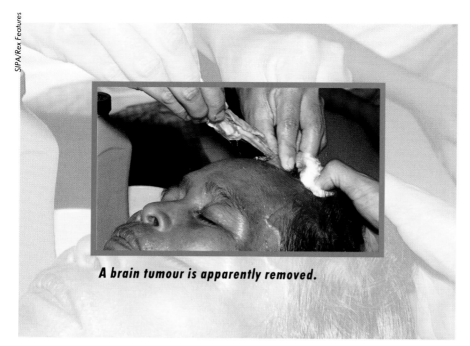

A brain tumour is apparently removed.

certain psychic surgeons were illustrated by a University of Kansas medical centre experiment in the 1950s. The test centred around a new surgical procedure designed to alleviate angina, the painful hardening of blood vessels in the heart. Not every patient was operated on – some received only a superficial incision giving them the illusion that they had undergone full surgery.

Even those who had not actually been treated often showed a significant improvement. The power of the mind, if convinced that certain rituals in which it believes are being performed, is clearly phenomenal!

All in the Mind

Psychic surgery does not guarantee a cure – but nor, for that matter, does conventional medicine. However, some who visit these unorthodox healers go as a last resort and return with reports of positive results.

Some practitioners have observed that working on westerners takes considerably longer than on their own people because those from developed countries are naturally suspicious. Are psychic surgeons charlatans or miracle workers? The answer, it seems, lies as much with the patient as it does with the healer.

ENTRANCING HEALERS

Three witnesses watched in amazement as Filipino healer Felisa Macanas materialized a piece of glass on 29th January 1975. This had allegedly been embedded in the thumb of a certain Alex Bull for eight years. The glass was not bloody, and Bull had no wound to show for his experience. All the healer had done was enter a mild trance and place two fingers of each hand above Bull's old injury while her husband held an open bible about a foot above her head. Bull had no further problems.

Such abilities are stock-in-trade for Filipino healers who supposedly base their art on the shamanic practices of local mountain tribes. Juan Blance can apparently make a flesh incision simply by holding an observer's thumb about 30 cm (12 inches) above the place to be cut. Alex Orbito and Marcelo Jainar can remove and replace eyeballs.

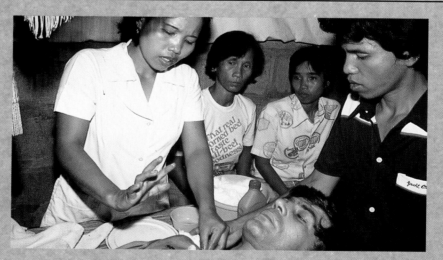

And Tony Agapoa extracts deeply embedded molars with his bare hands.

Most have the ability to materialize and dematerialize matter at will. Hands disappear into flesh or produce blood and tissue which may or may not belong to the patient. Surgical cotton seems to disappear into the skin, only to be removed on the next visit. Close-up

Some Filipino healers operate with only their bare hands.

film shows it vanishing mysteriously between the healer's fingers. Some investigators have seen evidence of fraud, but this does not always apply to every healer. According to scientist George Meek, such phenomena 'violate what science knows about the material universe'.

PSYCHICS

EXTRA-SENSORY PERCEPTION

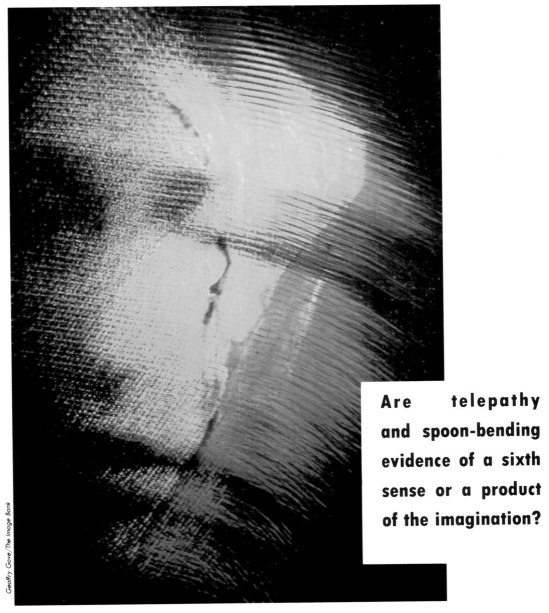

Are telepathy and spoon-bending evidence of a sixth sense or a product of the imagination?

Almost everyone has had what seems to be some form of psychic experience involving an element which cannot quite be explained. Perhaps you have thought of an old friend, when suddenly he phones you for the first time in months; or somehow you just 'knew' that something terrible had happened to someone very close to you.

Whether such incidents can be put down to coincidence or some form of extra-sensory perception remains unproven: but certainly there have been many highly intriguing cases which seem to point to the existence of some sort of additional or at least highly developed sense.

Psychic abilities have been shown to cover a wide spectrum, involving everything from tele-

THE UNEXPLAINED

Metal benders such as Uri Geller are able to warp spoons and forks simply by stroking them. Scientists are yet to prove that this really is mind over matter.

pathy to metal-bending as performed by Uri Geller. It is even thought that, as children, we are all natural psychics, but that scepticism quickly blots this power out.

The special abilities which psychics have take two forms: one is a type of extra-sensory perception, known as ESP, whereby they are able to receive information by unknown means. The second, 'psychokinesis', enables them to affect people or objects simply by using the power of the mind.

Although many psychics claim to have had the ability since birth, others sometimes become active spontaneously, following an accident, a near-death experience, or the loss of a loved one, revealing psychic abilities which have been dormant until this time.

Spiritualism

It is said that mediums, who receive messages from the 'spirits' of the deceased, have a special sensitivity which registers information not normally received by the five senses; and that this sensitivity can arrive quite suddenly. One mother, for instance, whose daughter died at the age of nine, had several visions in which a Jesus-like figure appeared. Just before the funeral, she was distressed to 'see' her daughter who 'told' her not to worry. The mother then contacted the local Spiritualist church and was taught to meditate and also to develop this psychic ability. She was soon able to contact 'spirits', and later became a professional medium.

Clairvoyance

Most mediums employ either clairvoyance or clairaudience. Clairvoyance literally means 'clear sight', from the French clair (clear) and voir (to see). It can mean seeing spirits objectively, or subjectively, when the spirit is visualized in the medium's head. Clairvoyance can

Doris Stokes, who described herself as a 'telephone exchange', put people in touch with deceased relatives by listening to their spirit voices.

AUTOMATIC WRITING

This mysterious phenomenon — whereby the pen, pencil or brush appears to move of its own volition — involves the production of letters, works of art or musical compositions that show a knowledge the writer could not possess on any conscious level. One Essex housewife, for instance, suddenly had a compulsion to paint, and for several months produced pictures in the styles of Picasso, Manet and Van Gogh, which far excelled those she normally painted.

Similarly, Rosemary Brown, who had only been taught the piano as a child, began to take down musical scores in the styles of Beethoven, Liszt and Mozart.

Interestingly, Mozart himself said that his music also arrived from somewhere outside himself and that he simply wrote it down. Many other musicians, writers and poets have described their creative processes as involving some form of 'dictation', too. The state of mind of the subjects varies widely: some are fully conscious of what is going on, while others go into a trance-like state and are totally unaware of the movements of their hands.

Psychologists explain automatic writing as a manifestation of the subconscious in which parts of the personality which normally remain hidden are revealed. Indeed, some psychotherapists even use automatic writing as a means of evoking deeply-buried memories.

▶ **Is automatic writing, which is often used by mediums, an expression of the subconscious, or do people actually receive messages from spirits or muses?**

◀ **Automatic painting often exhibits skills far exceeding that which the painter possessed previously. Many also have somewhat mystical themes.**

THE UNEXPLAINED

Channelling is a form of mediumship wherein the messages from the dead come via spirit guides who seems to 'take over' the medium and may even speak in their own voice and language.

CHANNELLING

Communication with discarnate beings who 'come through' via a channeller in the form of voices — often not his or her own — has aroused considerable interest over recent years, most particularly in the United States. Mysteriously, few channellers are themselves aware of what is said as the procedure appears to by-pass normal thought processes and does not enter the retrievable memory.

Whilst channelling, practitioners claim to be in touch with ancient prophets, deceased relatives, or even beings from other planets. Many bring spiritual messages, too, some of which are delivered with humour and in everyday parlance, whilst others use archaic words, strange accents, or even mysterious languages.

ing and two other monks had been hung over 400 years previously.

Dowsing

An ancient art, originally used to find water or lost objects, dowsing is still used today to find both water and oil, and to identify allergic substances and sources of geological stress. It can also be used to answer questions on almost any subject.

One method of dowsing to answer a question involves using a pendulum, comprising an object on the end of a chain. When the pendulum is held over a person's head it will rotate in either a clockwise or anti-clockwise direction to indicate 'yes' or 'no'. Which direction corresponds to a 'yes' or 'no' response depends on the individual dowser.

Practitioners also hold that lost objects can be located by holding a pendulum over a map or plan and moving it slowly until it gives a positive response.

Although a great deal of research has been carried out into the accuracy of dowsing, results have not been conclusive. But famous healer and psychic, Matthew Manning, puts it this way: 'To those who know, no explanation is needed; to those who do not know, no explanation is possible'.

also operate by means of subtle impressions received from the communicating spirits.

Clairaudience involves the hearing of spirit voices, and was used by Doris Stokes, probably the most famous medium of the 20th century. She often managed to convince sceptics by revealing a great deal about them, particularly trivial details that nobody else could know.

Psychometry

Another method of receiving information used by psychics is psychometry — the reading of objects, such as rings or watches, which have been worn by the subject. Through such objects, practitioners are able to ascertain the emotional state of the owners and to reveal details of their characters by 'reading' the vibrations given off by the object. Traumatic events in a subject's life also seem to register particularly strongly.

One psychic was given a very ordinary piece of stone to read, and instantly began to sob, later revealing that she had an upsettingly vivid picture of a monk being hauled up a hill to his death. It appeared that the stone actually came from the ancient site of Glastonbury Tor, in Somerset, where Abbot Whyt-

Dowsing with a pendulum can be used, like dowsing rods, to find something. This can be done by holding the pendulum over a map and waiting until it begins to move.

PSYCHICS

DIVINE
INSPIRATION

J.S. Duplessis: Mozart at the Clavichord/Louvre, Paris/Lauros-Giraudon/Bridgeman Art Library

What inspires some people to create great and enduring works of literature, music and art?

The idea that poets, like prophets, are simply vehicles or mediums for some divine voice that speaks through them is a well-established one. William Blake, for example, writing about his great prophetic poem, *Milton*, from which the line 'And did those feet in ancient times' is taken, declared with disconcerting sincerity, 'I have written this poem from immediate dictation, twelve or sometimes twenty or thirty lines at a time, without premeditation, and even against my will.'

To the ancient Celtic bards or the early Greeks, who worshipped Apollo and the Muses as the sources of musical and literary inspiration, the divine voice was a fact of life. In later times, young Greeks were taught the rules of harmony, metre and composition, but the ability to write great poetry was still seen

85

THE UNEXPLAINED

as a divine or god-given gift.

Those who denied this and claimed that it was just a question of their own talent were reminded of the legend of Thamyris, a Thracian bard, who, after boasting that he was the equal of the Muses, was struck blind and dumb and deprived of his powers of playing the lyre.

The Muses

The Muses were originally the nymphs who presided over sacred springs, the waters of which were said to have inspirational properties. The Pierian spring near Mount Olympus remained the principal place for their worship, even after they had been created goddesses.

Each of the nine Muses had her particular province and between them they covered the four major branches of poetry: epic, lyric, sacred and erotic; as well as music, history, tragedy, comedy and dance.

Differing Beliefs

The word 'inspiration' means literally 'breathing into'. It is a concept found in many cultures besides the Greco-Roman tradition from which we inherited our Western idea.

The Maoris say that the gods reveal their presence through a mystical wind, and religions throughout the world believe that a man can be inspired through various forms of possession, when a spirit or a god enters the body. The Christian religion is no exception. From the Day of Pentecost, when the Holy Spirit descended on the disciples in the form of 'tongues of fire', to the devotional poetry of the 16th-century Spanish mystics, St Teresa and St John of the Cross, the history of the church is filled with examples of inspired utterances and writings.

AUTISTIC CHILDREN

In the past the word 'autism' was used to describe any kind of withdrawn behaviour; today it is applied to one very specific mental disorder. The most noticeable symptoms in autistic children are that they appear cut off from the reality of the outside world and have great difficulty in learning to speak. For reasons that are not yet understood, their perception of the world is incomplete; they have a very limited idea of themselves and other people, which makes it almost impossible for them to form any emotional relationship. Usually they grow up severely retarded by normal standards, but have 'islands of ability', skills which they have mastered as well as any other person, and a few of them possess exceptional gifts.

These often involve prodigious feats of memory or high-speed calculation. One extraordinary ability, which has been described in several different individuals, is to give the correct day of the week for any date — in the past or far in the future — with hardly a moment's hesitation. The people who can do this are unable to say how much of their talent is due to memory and how much to calculation — they 'just know'.

In London, one young boy astonished artists and architects with his drawings from memory of buildings that he had looked at for no more than a few minutes. It is not just his recall of detail that is remarkable; the drawings are executed with a feeling for line and perspective that very few professionals could hope to emulate.

Another field where sufferers from autism have demonstrated a phenomenal 'photographic' memory is music, some being able to reproduce any piece on the piano after hearing it just once.

Stephen Wiltshire possesses a rare talent — after studying a building for a few minutes he can produce an accurate detailed drawing that impresses trained architects.

Whatever their feelings on religion, most European writers have never abandoned the idea of inspiration or the image of the Muses. Few writers would go so far as Blake and claim that their works are dictated to them by angels. However, nearly all writers admit that they have moments of inspiration when a line, an image or an idea 'just comes to them'.

Novelists have periods of intense creativity, when they feel 'possessed'. George Eliot said that, when she was producing her best work, some spirit, who was 'not herself' took over from her normal everyday personality. Other novelists have de-

Apollo, the Greek god of light, music and prophecy, had many functions, including the protection of flocks.

scribed how their characters have a life of their own. Dickens relates that, as he dozed in his chair, characters would enter his imagination, 'as it were, begging to be set down on paper'.

The Genius

What is it in great writers that allows them to draw on their store of words, memories, images and feelings to create something original and rich in content? The simplest answer is 'genius' – but what does this mean? The literal translation of the Latin word is 'creator' or 'begetter'. A man's genius was the spirit that brought him into the world and accompanied him through life, trying to influence his destiny for good.

The word was not used in its modern senses of 'having genius' or 'being a genius' before the 18th century. Since then, people who have tried to define it have

Christians believe that the Holy Spirit descended from heaven and entered into the disciples.

THE UNEXPLAINED

either seen it as an abnormal state of mind, akin to madness, or have dismissed it as 'one per cent inspiration and 99 per cent perspiration'.

Child Prodigies

Genius clearly takes many forms. Its most miraculous manifestation is when seen in a child prodigy. Music, mathematics and chess are the fields where youthful prodigies are most common, perhaps because they are abstract subjects with their own rules and logic. The precocity of Mozart, who started to compose at the age of four, gave recitals at five, toured Europe at six and had sets of sonatas published at seven, is the best-known example of all, because his mature genius lived up to the creative promise of his early years.

The Muses, companions of Apollo, were worshipped as sources of musical and literary inspiration.

VOICES FROM BEYOND

Some people claim to receive inspiration for their work or actions from beyond the grave.

This guidance may come from people they know who have died, or from total strangers, who have passed away many years before.

The phenomenon generally manifests itself through art, music or writing and is referred to as 'automatism'. In each case the person seems to be possessed by the spirit of the deceased to write, draw or play music in their particular style.

The famous British psychic, Matthew Manning, who was the centre of much poltergeist activity from an early age, realized he could write and draw under guidance from some other person when he was 16 years old. He even found himself writing in

Through automatic writing, Pearl Curran wrote novels dictated by a 17th-century Quaker girl.

foreign scripts such as Greek and Arabic, and would paint in the style of Goya and Picasso.

Tests carried out on Manning seem to suggest that he had redirected his poltergeist energies to more creative things. Sometimes the inspiration took a practical form. On one occasion he claimed that his grandfather had given him the winner of the Grand National, Red Rum, through automatic writing.

One automatic writer enjoyed a successful literary career. This was Mrs Pearl Curran of St Louis, who claimed to be in touch with Patience Worth, a young Quaker girl who emigrated to America in the 17th century. Patience was killed by Indians and apparently contacted Mrs Curran at her ouija board in 1913 by means of automatic writing.

PSYCHICS

THE PSYCHIC QUEST
C A S E B O O K

Fine Art Photographic Library

When Parasearch began a magical mystery tour of the Midlands, little did they realize what they would find...

Throughout history there are many accounts of mystics who discovered holy relics following a vision or a dream. One famous 15th-century saint, Joan of Arc, learned through mystical voices of the exact spot where her sword – for 'liberating' France – would be found. And in Tibet today, special monks are trained to search out psychic clues that lead them to sacred relics.

Psychic questing – the retrieval of ritual or religious objects by paranormal means – underwent a modern revival in Britain in the late 1970s when a team of researchers embarked on a mystical treasure hunt that took them all over the Midlands.

The Search Begins
It all began quietly enough in early 1979. Two Midlands-based parapsychologists, Andy Collins and Graham Phillips, set up a group called Parasearch and launched a magazine called *Strange Phenomena*. 'We were fed up with the way that years of psychical research had proved nothing,' said Collins.

Instead of trying to prove that the paranormal existed, they decided to collect information from British psychics and put it to practical use. To what purpose remained unclear for some months. Messages were duly received from all over the country but they seemed to be completely unrelated.

But in September, some of the stories developed a tenuous link. 'A number of the psychics began ringing up to say that they felt something significant was going

89

to happen,' said Collins. He and Graham waited in anticipation.

The Quest Revealed

One night when the pair were experimenting with hypnosis, Graham slipped into a trance. In the voice of a woman, he told the tale of an Egyptian pharaoh, Akhenaten. Apparently, his followers had been forced to flee to Britain after his death – to 'found a new colony'. During several sessions, the spirit guide revealed that Parasearch was in the position to find a long-lost talismanic stone which had originally belonged to the Egyptian refugees.

The amulet, she told them, had magical links. From the Egyptians, it had passed to a Celtic queen, then down through a diverse chain of people with occult connections – the Knights Templar, Mary Queen of Scots and the Gunpowder Plot conspirators who tried to kill King James I. Their only tenuous link was a common interest in a magical tradition which had been developed in Ancient Egypt. The group's quest would be a race against time – the stone must be found before the night of Halloween on October 31st!

For the next three weeks, the group were led a merry chase across the Midlands' countryside guided only by psychic clues. They followed – and were sometimes led astray by – dreams,

Owners of the mystic stone – Akhenaten, Mary Queen of Scots and Gunpowder Plot conspirators.

visions and clairvoyant information. 'Most of the time we thought we were going nowhere – that we'd find nothing,' said Collins. Nevertheless, they continued to follow up every lead. Then the improbable happened!

Found – a Sword

Following up one psychic snippet of information – that implied they would find a sword which would lead them to the talisman – several members of Parasearch went to investigate an old stone bridge at Knight's Pool near Worcester. 'I jokingly said that the sword would be concealed behind the ninth stone along and ninth stone down. When we got to that one there was a cavity with a blade of a sword laying across it,' said Collins. 'It was purely intuitive. Nine had already figured a few times in our quest.'

With renewed enthusiasm, the group resumed their search for the stone and turned to young Gaynor Sunderland – a 13-year-old UFO contactee and psychic from North Wales – for help. Accompanied by her mother, Andy and Graham, she returned to the Knight's Pool. There, using the sword as a divining rod, she led the others to a nearby ruin – Dunstall Castle.

Gaynor believed they would

Psychic Gaynor Sunderland returns to Knight's Pool where she uses the sword as a divining rod to lead the group to nearby Dunstall Castle.

THE SEVENTH SWORD

The ritual sword, discovered during the search for the green stone, is not the only one of its type to be found in psychic quests.

In all, six identical swords have been found by British questers in the last decade – five in the Midlands and another at Tintagel, the rumoured birthplace of King Arthur. Dated to the 19th century by experts, Andy Collins believes the swords are actually cast from moulds of an 18th-century ritual implement, which may have been used by Scottish Freemasons with a Knights Templar connection.

According to psychic information received by Collins, there may be a seventh sword. He discusses his theories in his book *The Seventh Sword*, published in 1991. He hopes that it will encourage questers to launch into the greatest psychic 'treasure hunt' ever known in Britain. 'I expect groups and individuals will start following up psychic leads all over the country,' he was reported as saying.

Treasure hunt: Andy Collins claims one more sword will be found.

find a clue at the top of the tower. But as she and Graham climbed the stairs, they heard the sound of 'giant flapping wings'. When dust and rubble began to fall, they fled in terror.

Around the Bend

Time was running out. On the evening of 29th October, members of Parasearch sat up late poring over books and maps. Suddenly Gaynor's mother remembered that mystical swans had featured heavily as one of Mary Queen of Scots' emblems. On the map, they located a bend in the River Avon called Swan's Neck near Birlingham in Worcestershire.

That night, Gaynor dreamed of a swan with a pouch around its neck. She intuitively knew it contained the stone. The very next morning, Parasearch made plans to go to Swan's Neck as soon as they could.

For reasons that he cannot explain to this day, Graham Phillips felt strangely compelled to drive to the river bend alone. On arrival, he immediately recognized a tree-lined avenue that had featured in one psychic's vision.

Mary Queen of Scots' tapestry led the group to Swan's Neck where an avenue of trees held the clue to the resting place of the green stone.

Then he stumbled over a grassy mound. Twice he dug down into the earth with a trowel – but found nothing. On the third attempt, he uncovered a small casket. Inside was a stone in two-tone shades of pale green, shaped like half an egg.

'For months we just sat around and looked at the sword and the stone,' said Collins, 'and wondered what we were supposed to

THE UNEXPLAINED

do with them'. He believes the 19th-century sword is a Scottish Freemason emblem of office and the stone a symbolic relic of the Egyptian magical tradition. But many questions remain unanswered about the green stone. If it really is Egyptian how did it come to England? There is little more than circumstantial evidence to suggest that Akhenaten's followers ever travelled so far, although Collins has his own ideas, apparently based on psychic information.

Perhaps in the future more light will be thrown on this strange affair but, for the moment, any concrete answer remains buried in hundreds of years of uncharted history.

New Quests

Andy Collins, however, has been spurred on to embark on further quests. Working with other psychics, he has since found a number of sacred sites and relics scattered throughout Britain. According to him, any reasonably sensitive person can tune into the 'message' which is given off by these objects.

In this sense, questing is a two-way process, in which the quester is both a hunter seeking his quarry and a passive figure, drawn to the goal by the power of the object itself.

This active participation contrasts sharply with the traditionally passive role of the medium. As Collins puts it: 'It takes the psychic out of the seance room and into the landscape.'

THE GRAIL QUEST

One of Andy Collins' most intriguing quests was centred around the Glastonbury Zodiac.

This earth zodiac in Somerset was first written about in the 16th century by the famous occultist John Dee, but not traced out until the 1920s. Natural land features such as rivers and hills, along with man-made earthworks form the figures of the 12 signs of the zodiac.

The area has strong Arthurian associations. Legend has it that Joseph of Arimathea set up the first British church at Glastonbury. It is also reputedly the burial place of King Arthur.

For modern-day questers, these associations are significant. The Holy Grail – the chalice Christ used at the Last Supper – was reputedly brought to England by Joseph of Arimathea and later became the subject of the famous quest for the Grail by King Arthur's knights. The ten-mile wide circle of the Glastonbury Zodiac is said to correspond to the Round Table of Arthur's Court at Camelot. For example, the sign Sagittarius is a knight on horseback – perhaps Arthur himself. Other sources link the Grail with the Chalice Well, located in the sign of Aquarius. While working his way round the Zodiac in 1985, Collins uncovered an 18th-century ebony cross at Wimble Toot, in the sign of Virgo.

Glastonbury Zodiac's Sagittarius (left) and Chalice Well (below).

PSYCHICS

WHAT KATIE DID!
CASEBOOK

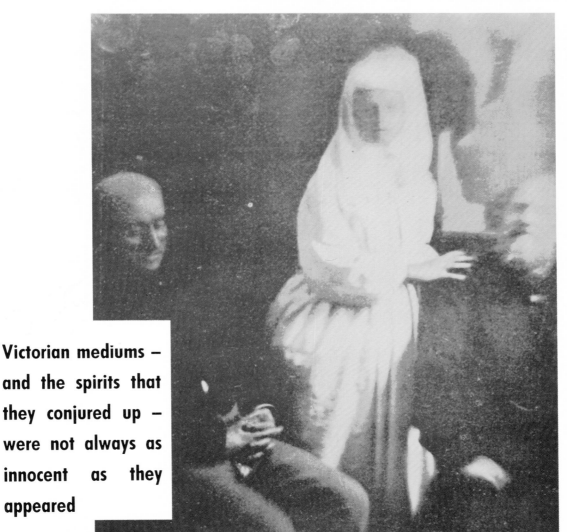

Victorian mediums – and the spirits that they conjured up – were not always as innocent as they appeared

In a darkened 'cabinet' sectioned off from the main room, a young Spiritualist medium lay bound to a chair, slumped in a trance. On the other side of the partition her audience – the 'sitters' – waited patiently in the gloom until, with a slight rustle, the dim figure of a girl dressed in white emerged to walk among them.

The medium was Florence Cook, an attractive 17-year-old from London's East End, and this was one of many seances at which she demonstrated her full-form spirit materialization. But on this occasion, the 'spirit' girl had hardly taken a few steps when one of the sitters, a Mr Volckman, leapt up and grabbed her round the waist. The others, outraged, fought him off; in the confusion he lost some whiskers and the gaslight was shut down.

The spirit speedily retreated into the cabinet, and several minutes passed before it was opened to reveal the medium – apparently still in a trance. Although Florrie Cook was flustered and her hair was coming unpinned, the tapes that bound her were still intact. Had she performed a feat of escapology? Or was the girl in white a genuine visitor from the spiritual plane?

When she began practising as a medium in the early 1870s, Florrie's talents created a stir from the beginning. Seances that centred around her were hardly prim and proper, for the spirits often threw her in the air and ripped all her clothes off, ensur-

93

ing no lack of sitters for subsequent meetings!

But it was the emergence of the materialized spirit called Katie King that brought believers flocking to her door, for Katie was no ethereal being – as many a gentleman sitter found out to his delight. Katie would perch on their laps to prove she was flesh and blood – if only temporarily – and sometimes invited them to pinch and stroke her.

Katie King was the 'spirit' name of the daughter of the 16th-century buccaneer, Sir Henry Morgan. The 'King' family seemed to be in demand as spirit controls; some six years earlier Katie herself had been the star materialization at seances held in the USA by the Davenport brothers, only to shift her allegiance to Florrie in 1874.

Katie's conversation was hardly elevating, consisting mainly of small talk delivered in a whining Cockney accent noticeably similar to that of her medium, Florrie. Interestingly, when she had appeared at the Davenport brothers' seances, Katie's accent had been distinctly American.

Not surprisingly, there were many sceptics who claimed that Florrie and Katie were one and the same. It is also worth noting that Florrie's assailant, Volckman, was a close friend of Mrs Guppy – another famous medium, known to be jealous of younger rivals.

Scientific Proof?

After the scuffle with Volckman, Florrie became anxious to gain the seal of respectability. At the time she was being paid a substantial retainer by William Blackburn, a wealthy old man with a fondness for pretty spirits.

Now she turned up uninvited on the doorstep of the eminent scientist William (later Sir William) Crookes. Florrie knew that Crookes had just spent several months investigating the American medium Daniel Dunglas

The spirit control 'Katie King' – she bears a startling resemblance to Cook.

PSYCHICS

One of the few photographs taken of Florence Cook and 'Katie King' together, Cook slumped in a trance.

Florence Cook in 1874 – medium extraordinary or opportunist fraud?

SITTING PRETTY

Religious beliefs have long provided a cover for the release of repressed emotions, and the Victorian era was no exception. The 1870s saw the high point of mediumship as a form of popular entertainment, and it was perhaps small wonder that gentlemen flocked to the Spiritualist sittings of young ladies like Florence Cook and Mary Showers. Under the pretext of contacting the spirit world – what could be more moral – sitters had ample opportunity to feel the bodies of young and attractive materializations.

One feminist writer on the subject has questioned the passive role of the medium. In any other context, a girl bound to a chair and locked in a cabinet would raise strong implications of bondage fetishism. Were the pretty sitters of the Victorian parlour seance unwittingly part of an exercise with undertones of male sexual domination?

Home and declared himself impressed. Could he not do the same for her reputation?

Crookes went further. He was so taken by Florrie and Katie that he invited Florrie and her mother to stay at his house in Camden Town, North London, so that he could study the materialization in detail. He then turned his laboratory into a seance room and invited hand-picked friends and colleagues to see Katie as a materialized spirit. His wife was expecting their tenth child and was confined to her room for most of the time, so took no part in the experiments.

The scientist was convinced that the two girls were distinctly separate entities, and photographed them to prove his point. Unfortunately, of the 40 or so original plates, few have survived; Crookes destroyed many of them before his death in 1916, and those which remain provide insubstantial proof.

Nevertheless, several sitters recorded positive impressions of their encounters with Florrie and Katie. They noted physical differences in height and hair colouring, and that Florrie's ears were pierced while Katie's were not. Similarly, at one seance Florrie had a large blister on her neck, while the spirit's neck could be seen to be unblemished.

On at least one occasion Crookes claimed to have witnessed an extraordinary double act – Florrie arm-in-arm with her friend, the medium Mary Showers, while their respective materializations also walked arm-in-arm. Unfortunately, Showers later confessed to being a fraud – so what was Katie doing consorting with a fake? Inevitably, rumours spread that Crookes' experiments were no more than a cover for sexual intrigue.

The End of 'Katie King'

After three months of living in Crookes' home, Florrie revealed that she had secretly married a sailor named Corner a few months earlier. Coincidentally, Katie announced her intention to return to the spirit world. The experiment was over.

Following Katie's departure, Florrie's career as a medium ran into trouble. Her new spirit

THE UNEXPLAINED

materialization – 'Marie' – was seized during a seance and proved to be none other than Florrie in her underwear. The discredited medium retired shortly after this incident.

Today Florrie and Katie have few admirers among psychical researchers, although it has been noted that since Florrie was under such pressure to perform it would hardly have been surprising had she resorted to trickery when her powers failed her.

The New 'Katie King'

Despite all the arguments, the story has a highly provocative postscript. In 1974 a spirit calling herself Katie King materialized at the seance of medium Fulvio Rendhell – 100 years after the Crookes-Cook collaboration.

This time the 'spirit' was captured by a movie camera on infra-red film and appeared to be genuinely paranormal – although the 'new' Katie looked nothing like the figure photographed by Crookes. Whatever the truth, it seems unlikely that we have heard the last of Katie King.

Florence Cook exposed: the spirit control 'Marie' was discovered by Sir George Sitwell to be Cook herself.

'The Kiss' by Gustav Klimt – symbol of flirtation or submission?

Gustav Klimt (1862–1918)/The Kiss/Osterreichische Galerie, Vienna/Bridgeman Art Library

THE POWER OF THE KISS

The kiss, long seen as a profoundly intimate and potent act, can also symbolize submission – witness the kiss traditionally demanded of their subjects by British monarchs during the Coronation service. (In the rehearsal for the Queen's Coronation, Prince Philip gave his wife no more than a flirtatious peck on the cheek. 'Come back and do it properly,' was the Queen's sharp response.)

Similarly, while kisses can be a demonstration of great tenderness between lovers, there are many occasions throughout history in which they have been perverted into a symbol of evil. Judas betrayed Jesus to the Romans with a kiss, only to be so overcome with remorse that he hanged himself. Likewise, the kiss of a vampire spelt physical and spiritual doom for the victim, who would go on to suffer the same terrible fate as their demon lover.

Strangely, vampire tales – a heady mixture of all-powerful lovers, passive victims and lost innocence cloaked in a shroud of necrophilia – had many of the same pseudo-erotic overtones as the seances of Florence Cook. Perhaps it was no coincidence that, thanks to Bram Stoker's classic tale *Dracula*, they were popular at around the same time.

NATURE'S MYSTERIES

MARK OF THE BEAST

Mystery animals have left their paw prints on myth and legend. But do some of them still exist?

In June 1983, the sleepy town of South Molton on Exmoor's southern fringes declared war on a savage beast of prey.

A series of vicious attacks on sheep and lambs had begun five months earlier – and by June the death toll numbered 200 animals. In every case, the predator had felled its victim with a powerful bite through the neck, ripped the soft underbelly open and licked the bones clean.

Worried farmers formed armed patrols – each 40 men strong – to hunt the culprit. Sharpshooters from the Royal Marines were called in to help and a police helicopter was kept on constant alert. But it was all to no avail – the creature is still at large!

Jungle Scream

Eyewitness accounts of the animal vary from descriptions of a 'stone-coloured puma' to an

97

THE UNEXPLAINED

◀ Even camouflaged Royal Marines were unable to trap the 'Beast of Exmoor' whose favourite method of killing livestock was by a bite to the neck (inset).

▼ Is this the Beast? This photograph was taken by Exmoor naturalist Trevor Beer who claims it stands at least half a metre (two feet) at the shoulder.

'enormous black cat with a tail as long as its body'. 'The Beast of Exmoor' – as locals now call it – can also apparently run at speeds of 35mph, leap hedges and five-barred gates, and has an eerie nocturnal cry described by some as a 'jungle scream'.

Farmer Eric Ley lost £2,000 worth of livestock during the first year of the attacks. Like many other local people, he is convinced of the creature's existence and its 'supernatural' intelligence. How else could it have killed a lamb in a field guarded by the Royal Marines?

Local Police Sergeant David Goodman was stationed at South Molton when the attacks began. He believes the cat still strikes, quoting sporadic livestock deaths which bear its mark. Many more cases, he suspects, never come to light because the farmers are now 'used to living with it.'

A £1,000 reward offered by the *Daily Express* for the first unpublished photograph of the beast has never been claimed. Clearly, the mystery is far from being solved.

Big Cats

One killer animal roaming the countryside sounds bad enough, but many more exotic felines have been spotted all over Britain. There are a Cuffley Lioness and a Nottingham Lion – just to name two. And such stories are by no means only a late 20th-century phenomenon.

William Cobbett, the agrarian reformer, described in his 1830 book *Rural Rides* how he saw a huge cat crawl into a hollow tree in the grounds of Waverley Abbey near Farnham, Surrey. A boy at the time, he was beaten for lying.

Interestingly, a more contemporary mystery cat – known as the 'Surrey puma' – has been spotted in much the same area as Cobbett's feline. First seen in the late 1950s, its first reported killing in August 1964 quickly made the headlines. Just over the Surrey borders in Hampshire, farm manager Edward Blanks found a 16kg (35lb) calf carcass, which had been dragged on to his property from a neighbouring farm. Three days later, a bull weighing 205kg (450lb) was also discovered – seriously mauled.

Many witnesses who came forward agreed that the animal was a light tawny colour with a dark stripe down its back and a long tail. It had the head of a cat, stocky legs and large paws.

The Surrey police issued a public warning, and attempted to catch the beast themselves. Although their 'puma book' registered hundreds of sightings, the animal eluded them and now seems to have vanished.

One farmer who had better luck was Ted Noble, from Glen

NATURE'S MYSTERIES

THE BUNGAY BLACK DOG

According to a small booklet published by the Reverend Abraham Fleming, 'a straunge and terrible wunder' occurred during the morning service in the parish church of Bungay, Suffolk on 4th August 1577.

During a violent thunderstorm, the congregation was amazed to see an enormous black dog, its body flashing fire, enter the church and run swiftly through the building. Two people's necks were wrung — 'at one instant clene backward' — and another man survived, although shrivelled like 'a peece of lether scorched in a hot fire'.

The dog apparently also visited Blythburgh church, nearly ten miles away, where it killed three men and scorched many worshippers. Burns — reputedly left by the manifestation — can still be seen on one of the church doors today.

Phantom black dogs often figure in British and Irish folklore. Many have names — the Shuckey Dog haunts East Anglia and Skryker (from the Old Norse word for howling) lurks in the North. Some protect lone travellers; others are seen as an omen of death.

Legend has it that the dogs, known for their blazing eyes, can scorch, electrocute, and pass through solid objects before exploding violently. Ball lightning is capable of similar feats, and also produces the sulphurous odour often described during these phantom sightings. Might these two phenomena be linked in some way?

Black dogs tend to be seen near prehistoric sites, bridges, streams and churches. Those who believe such places store powerful earth energy claim the dogs could be ghosts of real animals set by our ancestors to guard sacred sites. But Kaledon Naddair, a present day Pictish shaman from North Scotland, has another theory. He believes that black dogs are just one of the many types of faerie-wildfolk who still live among us.

▼ *Visitation from a black dog. Blythburgh's church door (inset) still bears scorch marks from its phantom visitor.*

Cannich in Invernesshire. After losing many sheep to an unknown predator, the plucky Scotsman set his own trap in October 1980 with a sheep's head. One morning he made his usual routine check — and heard growling. He had caught a live puma.

Scientific analysis of the animal's droppings established that it had been living in the wild for quite some time. The riddle of how an native American big cat came to be roaming Scotland was only solved when a man confessed to releasing two pumas into the countryside before being sent to jail. However, this does not explain subsequent sightings of big cats in other parts of Scotland right up to the end of the following year.

Fact or Fantasy

Are Britain's strange big cats all descended from former pets or escaped circus and zoo animals which have bred in the wild? It is certainly possible that a number of them were once kept in captivity — particularly as the keeping of exotic felines was a popular pastime a few years back.

But this does not explain how most of them have managed to elude massive police searches for so long — often literally disappearing into thin air. For the farmers who have lost livestock, however, there is no question of fantasy. At the height of its reign of terror, the Beast of Exmoor gorged itself on 16kg (35lb) of meat — enough

A possible sighting of the elusive 'Surrey puma' taken from a distance of 32 metres (105 feet).

THE UNEXPLAINED

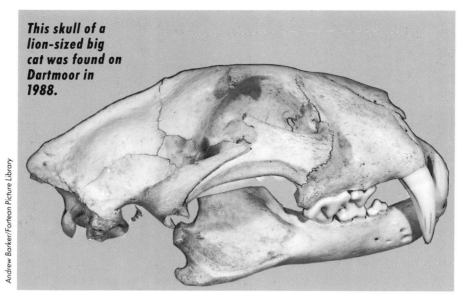

This skull of a lion-sized big cat was found on Dartmoor in 1988.

to feed 70 people!

The answer is further complicated by the fact that many witnesses have difficulty in deciding whether they have seen cats or dogs. Indeed, when tracks are found they usually show claws and as all big cats (with the exception of the cheetah) walk with their claws retracted, some have argued that the cats may well be dogs! The discovery of several dog-sized black cats (in the Moray region in Scotland during 1983–1985) with extended claws has confused the issue even more. Some researchers maintain that a new strain of cat – perhaps a cross between an abandoned domestic cat and a puma has come into being.

Blurred Boundaries

But so far the evidence is inconclusive and does not explain the other mystery beasts that have been spotted on British shores. Phantom dogs regularly appear and disappear; North American raccoons have been found in Northampton and North Wales; porcupines in Okehampton, Devon; not to mention crocodiles basking on the banks of the Stour near Sandwich in Kent and Arctic foxes frolicking in West Yorkshire.

The list, it seems, just goes on endlessly and the boundaries between fact and fantasy, for the moment at least, remain blurred.

THE BEAST OF GÉVAUDAN

Mystery beasts are not unique to Britain. The region of Gévaudan in south-east France also had its own terrifying creature.

In 1764, a 'fanged monster' with a taste for human blood terrorised the villagers of the town of Malzieu. Many died savage deaths and tales circulated about the beast's apparent invincibility. Panic set in and King Louis XV sent his personal gun carrier, Antoine de Beauterne, to rout the beast in July 1765.

A true huntsman, Beauterne spent three months checking the lay of the land before taking action. One September morning, he set out with 40 hunters and 12 dogs. They encircled the ravine where the beast was supposed to lurk – and waited.

Suddenly, it careered into the clearing. Beauterne took aim, a shot rang out, the animal fell but rose again and made an attempt to escape. It collapsed dead on the third shot. The skin was stuffed and the beast – a rare species of wolf measuring two metres (six feet six inches) from head to tail and weighing 65kg (143lb) – was brought in triumph to the king's court.

But the attacks continued and another 14 people died brutally in the spring of 1767. Local hunter Jean Chastel, accompanied by 300 men, set out in June to kill the foe. With a bullet blessed by a priest, Jean himself shot another wolf. The killings ceased for good.

NATURE'S MYSTERIES

MYTHICAL CREATURES

Will we ever solve the mysteries of the monsters that inhabit the earth?

Animals that are rarely seen, almost never caught, and defy definition have long been an intriguing puzzle worldwide.

A classic example is the Loch Ness monster. Seen by thousands of people and convincingly photographed, its existence continues to be a point of lively debate. The Siberian equivalent, the monster of Labinkir, is said to lurk in the freezing waters of Lake Labinkir. Although many sightings have been reported, Soviet biologists have failed to get even a fleeting glimpse and have concluded that it is probably a giant *schuka* fish, which can reach 2.7 metres (nine feet) in length.

The Swedish authorities take the existence of their famous lake monster, Storjooduret, reputed to be a huge serpentine creature measuring up to 20 metres (65 feet) long living in Lake Storsjön, much more seriously. There have been many expeditions to find it and it was recently declared that anyone trying to capture or kill the creature could face prosecution.

Chinese scientists also believe in the existence of ape-like creatures, known as the wild men of

101

THE UNEXPLAINED

Hupeh. In 1939 a peasant woman disappeared for 27 days, claiming on her return that several of these beings had abducted her and taken her to a forest. Nine months later she gave birth to a simian-looking baby who survived into adulthood and died in 1960. In 1980 his bones were dug up and examined by Chinese scientists, who confirmed that they had both ape and human characteristics. They have now created a special reserve in the national park for the wild men.

Ape-like beings like these, (sometimes referred to as BHMs or Big Hairy Monsters), have been reported in the USA, Russia, Canada, British Columbia, Australia and even Britain.

The Mysterious Yeti

In the early 1920s Tom Crowly, a seasoned climber, was coming down a mountain in Glen Eanaich, Scotland, when he heard footsteps behind him. He turned to see a huge grey figure with pointed ears, long legs and feet with toes like talons.

The most famous Big Hairy Monster, however, is the legendary Yeti or Abominable Snowman, who is said to inhabit the Himalayan mountains. There have been many more sightings of possible footprints

The Loch Ness monster (inset) as captured on film from Urquhart Castle in May 1977.

▲ **On 30th November 1861 the crew of a French gunboat sighted a giant squid off the coast of Tenerife. However, after a struggle all they managed to catch was its tail.**

NATURE'S MYSTERIES

than of the controversial creature himself, but even face-to-face confrontations have done little to convince disbelievers.

In 1986 Anthony Wooldridge was on a solo charity run in the Himalayas and found himself staring across an expanse of snow at a large erect figure standing behind a shrub. It was about two metres (six feet six inches) tall with a large square-shaped head and the whole of its body appeared to be covered in dark hair. He whipped out his camera and took the first-ever photographs of the mysterious Yeti.

Two frames from film footage of the famous 'hairy monster' Bigfoot, taken in 1967 in California.

ANIMAL GHOSTS

In 1916 Arthur Springer, a retired CID inspector from Scotland Yard, photographed two ladies taking tea in a summer garden. To their utter astonishment, when the plates were developed the figure of a headless dog was discovered standing beside them.

During the 1890s a large white cat was often sighted along the banks of the Trent in Lincolnshire, terrifying everyone who saw it. One snowy night a local farmer kicked it as it streaked across his path. The next morning the man found no evidence of paw marks, only his boot prints.

Many people believe these ghosts to be manifestations of the devil, although some say that they are the spirits of animals that are linked to ley lines, prehistoric tracks where the earth's energy enables them to appear.

The ghost of a headless dog photographed by Arthur Springer.

THE UNEXPLAINED

One of the several winged cats which have been reported in Britain.

The zoologist Desmond Morris has said the creature could be one of three things: a hitherto unknown species of bear or giant monkey, a holy man or hermit, or the fabled Abominable Snowman himself. Professor John Napier, a leading authority on primates, said the photographs had reversed his scepticism.

However, it seems that no matter how many times a mysterious animal is sighted or even photographed, the scientific establishment needs a corpse before conceding its existence.

Sometimes even a corpse is not enough. In 1966 a 'flying cat' was shot down near Ottawa, Canada. Scientists are still undecided as to whether the protrusions on its back were just two oddly positioned tails or whether they really were wings.

Bloodthirsty Werewolves

Perhaps the most mysterious animal of all, however, is the werewolf. Stories of men capable of transforming themselves into bloodthirsty wolves have been told for thousands of years. In the Middle Ages, many thousands of people were burned at the stake for supposedly being werewolves, and attacks are still reported in newspapers worldwide today.

However, the evidence which appears to 'prove' the existence of mysterious animals has not solved the great mystery. It only seems to rekindle serious interest among experts as to whether they really do exist.

ANIMAL HEALERS

The important role played by animals in healing was first recognized by the medical profession in the 1970s. Doctors noticed that people who kept pets had a greater resistance to some illnesses and that rhythmical stroking of their animals actually lowered blood pressure.

Since then, cats in particular have been used for many types of therapy. There are animal healers that are reputed to have powers equal to their human counterparts. One of the most famous was Rogan, a marmalade cat rescued from the Cat Protection League in 1971. Rogan was said to be able to cure anything, from slipped discs to fading vision, simply by laying his paws on the sufferer on several sessions.

More recently, a dog was thought to have diagnosed her owner's cancer by trying to bite off a mole from her leg. She became so persistent that the woman visited the doctor, whereupon the cancerous mole was immediately removed. Research on the subject then began.

It has been found that stroking pets lowers blood pressure.

NATURE'S MYSTERIES

The mysteries of
SHAPE-SHIFTING

The desire to alter shape sometimes extends to an identity change, whereby humans take on another form

Many apparent miracles centred on the strange 19th-century medium, Daniel Dunglas Home. Among the most remarkable was his growth in height by a full 15 cm while in a state of trance – something he was seemingly able to perform at will. This feat completely perplexed anyone who witnessed it. Yet claims about changes in bodily shape were nothing new: for centuries, accounts of humans who could actually take on another form altogether – that of an animal or bird, perhaps – were very much part of tradition in many areas of the world. So it is perhaps not surprising that a great deal of folklore has grown up surrounding these tales.

Werewolf Wizardry
One popular belief, common until quite recently in remote parts of Eastern Europe, involved werewolves. These howling, grey-coloured creatures with slanting green eyes were supposedly human by day but animal on nights of the full moon.

This terrifying phenomenon, known as lycanthropy, has often been explained away by psychoanalysts as a type of mental disturbance, particularly in view of the animal-like sounds which some psychotics are prone to make. It has also been thought to occur as a hallucination during a session of self-hypnosis.

Those who claimed contact with werewolves commonly be-

THE UNEXPLAINED

Donning the hide of their animal familiar is a ritual practice among tribal peoples all over the world.

lieved these beings could be recognized by certain outward signs in their human form. In Greece, for instance, eyebrows meeting above the nose, hairiness, claw-like finger nails and small, pointed ears were all said to be sure indications. It was feared, too, that if you became a werewolf, the affliction would last nine years – unless you partook of human flesh during this time, in which case you would remain a werewolf forever.

What is a Werewolf?

The condition of lycanthropy was sometimes said to be hereditary, or to occur if you had broken a taboo. Some believed that werewolves were simply hysterical individuals dressed up in animal skins or under the influence of hallucinogenic drugs (possibly belladonna, known to induce a sensation of having hairy skin).

Others were convinced that werewolves were victims of enchantment, and that a witch's spell could only be broken if you pointed at the human form, while shouting the accusation 'You are the wolf!' or calling out his or her Christian name three times. However, there was always the risk that if you did this, the spell might be transferred and you might yourself end up enveloped in the guise of a wolf.

Wild-beast Counterparts

In Africa, too, there are many similar accounts of humans who at times will take on the form of an animal – a hyena, wild boar, vulture or crocodile, for instance. In some tribes, the

THE TOTEM TRADITION

When a woman of the Zapotec tribe of Central America is in labour and about to give birth, all the relatives get together and proceed to sketch a series of different animals, erasing each outline as soon as it has been made. The drawing that is being prepared at the moment of the birth has a particular significance, however, for this very creature is destined to become the baby's totem, housing one of his or her souls. Once old enough, the child will be given an animal of that species to tend personally; and progress through life will be continually monitored alongside that of the animal. Indeed, some people think the two are destined to die together.

Such animal or plant totems are found among many peoples of the world, providing them with confidence through a suggested link with the spirit world. Certain clans in Sumatra, for instance, believe that they are each descended from a particular creature, to whom they have obligations, including a ban on eating the meat of their totem, be it wolf, bear, buffalo, cat or tiger. The totem connection is made during complex male initiation rites. These involve a mock ritual killing of the adolescent boy and symbolic transference of his soul to that of the clan's totem, and vice versa. Thereafter, he takes on the name of the creature and regards all that species as his brethren.

Many of the Aborigines of Australia also traditionally believe that they have animal familiars, as do North American Indian tribes who take totems as personal as well as group protectors.

The traditional totem pole of American Indian tribes represents their animal or bird protector.

initiation of a shaman or witch doctor still generally involves the mingling of his blood with that of the animal.

An intimate union between human and animal is then thought to result and the shaman believes he has supreme power over the animal, to the extent that he is convinced the creature will act on his behalf and even kill if so commanded.

Members of some African secret societies are also known, at the instigation of their ju-ju man (shaman), to dress in the skins of wild creatures and to attack and eat human victims.

Hippo-men

Such beliefs can give rise to bizarre situations. The British periodical, *The Listener* (21st August 1947), reported the case of a West African villager who took his neighbour to court after a hippopotamus damaged his cabbage patch. This happened because the neighbour belonged to a local secret society that believed in the phenomenon of animal counterparts, and he believed his own counterpart or 'bush soul' was a hippo. In his defence, he pleaded that he had instructed the hippo not to wander over other people's land and cause damage or it might be shot. Apparently, however, the hippo had disobeyed and the judge was adamant that those who have hippopotami as their bush souls must control them.

Animal Familiars

In Siberia, the Yakut tribe maintains an ancient tradition that a magician keeps a second soul in an animal seen only by him. The most powerful magicians house their additional souls in bears or eagles; the weaker, in dogs.

A Calabar tribe, living by the River Niger in Africa, believes that each male has four souls, and that one is permanently hidden in a forest animal. Once the tribal magician has identified the beast to him, the individual will do everything not to harm any creature of that species, for fear that he may die or be injured, too. Similarly, if the man should die first, his animal familiar is said to perish in a frenzy.

Transference of Souls

Deceased members of a tribe often have wounds identical to those of their animal counterparts. The sceptical will, of course, argue strongly that these injuries were probably self-inflicted, after the person heard about the animal's death. But there are many parallel incidents that have been reported the world over. An 18th-century Scottish trial, for instance, involved one William Montgom-

The pooka is an Irish hobgoblin who appears to mortals in the guise of a horse or donkey.

The Greek deity Zeus changed himself into many different animals in order to seduce young girls – in this case, Europa.

THE UNEXPLAINED

ery who confessed to killing two howling cats which he thought were witches in animal form. A short while after, two women, who had been suspected of witchcraft and who lived nearby, died suddenly.

The transference of souls is believed to come about in various ways. The Taman tribe of Burma, for example, are convinced that the gods can enable them to enter the body of a tiger if they roll about in earth on which the animal has urinated. Such is the strength of this belief that, throughout the period for which a Taman's soul is thought to be in the body of a particular tiger, the human form appears deathly pale and breathing is very shallow. Once out of his trance, so anthropologists recount, the tiger-man will provide a most vivid description of all the activities he has enjoyed while in his animal form – mating and killing among them.

Mischief-makers

Not all occurrences involving shape-shifters are so violent, however. In Japan, the *tanuki* – a prankster who gets up to all manner of practical jokes and non-malicious tricks – is said to resemble a badger and to be capable of changing itself into a man or some kind of creature, or even an inanimate object.

The *pooka* of Ireland generally has the form of a horse or donkey, and is thought to become human at times, helping those he meets to understand the language of the animal world. But the *ovinnik* of Eastern Europe, which looks like a black cat and barks like a dog, is an ill-humoured shape-shifter reputed to set fire to barns; and the *ku* of Hawaii is a large hound, also deemed to have the magical power to change form and to crave the meat of man.

TALES OF TRANSFORMATION

Stories involving external souls that enter other creatures, animal counterparts and changing forms have had a great deal of influence since early times. The ancient Greeks, for example, had many a tale of the gods descending to earth in a different guise to seduce a mere mortal. There are many such legends worldwide, like the familiar *Beauty and the Beast*, that reflect these ancient beliefs.

In Japan, numerous stories about change in physical form are thought to be the remains of a shamanistic cult that once flourished there. In one example, a handsome young fisherman catches a magnificent coloured turtle which vanishes by morning. But in its place is a most beautiful woman who lures him to a fantastic palace several fathoms under the sea. Here they marry, but after three years of wedded bliss, the fisherman becomes homesick and begs for a chance to see his family once more. Taking with him a box which he is told never to open if he wishes to return to life under the sea, he finds a supernatural time-lapse of 300 years has changed everything he remembered on land. In despair, he opens the box, an action which results in disastrous physical change: his hair turns white, his face is suddenly etched with wrinkles and finally he crumbles into dust.

In the famous fairy tale, Beauty and the Beast, *the beast reverts to human form only after Beauty has kissed him.*

When Apollo tried to seduce Daphne she appealed to the Earth goddess, Gaea, who swallowed her up and left a laurel tree in her place.

108

NATURE'S MYSTERIES

WISDOM FROM THE DEEP

Are the games that dolphins play part of their attempt to communicate with us?

It is not only the alert glimmer in dolphins' eyes that inspires questions about their intelligence. Their broad, engaging smiles suggest a capacity for humour, and the way they approach people without a hint of fear has intimated to some that perhaps dolphins have something to tell us. Scientists are still speculating as to what this might be.

Aquatic mammals, dolphins swim in schools of up to 1,000 and research has shown they have a strong sense of brotherhood. A group will help an injured dolphin remain buoyant for as long as four days to ensure that it can breathe. Environmental pressure group Greenpeace has also reported that dolphins help others trapped in fishing nets – often at great cost to themselves.

Underwater Ethics
This 'social conscience' is sometimes extended to man himself. Indeed, stories of life-saving dolphins date back to ancient times.

In the fifth century BC, Greek

THE UNEXPLAINED

Age-old fascination – Minoan frescos dating to 1400 BC.

historian Herodotus told the tale of Arion, a rich and famous musician, who set sail for Corinth. Sailors on board the ship hatched a murder plot to relieve him of his wealth but he learned of their plans and begged for his life.

Faced with the choice of either killing himself or throwing himself overboard, Arion decided to risk the waves. As the musician floundered in the water, a dolphin mysteriously appeared at his side and transported him safely back to the shores of a town called Taernarum. Happily, the sailors were later brought to justice.

Some choose to write off this story as mere myth, but to this day, a small bronze statue stands in Taernarum at a place known as Arion's Temple. It portrays a man astride a dolphin.

Sea Saviour

Tales of life-saving dolphins still circulate today. In the Cocos Islands off the coast of Thailand, 11-year-old Nick Christides was surfing in 1982 when he was carried right out into the Indian Ocean by a strong current. Try as he might, he could not swim back to shore.

For the next four hours, Nick drifted in the dangerous shark-infested waters while boats and planes searched for him in vain.

Dingle Bay dolphins befriend visitors to the Irish west coast.

Suddenly, a dolphin appeared at his side. It fended off several circling sharks, then swam close to the child making sure that he did not drown.

Eventually, Nick was spotted and brought safely to shore where he told reporters of the aquatic friend that had saved his life by swimming round him in circles throughout his ordeal. Nick remained firmly convinced that the dolphin came to his assistance because it realized how dangerous the current was for a small boy.

Bright Sparks?

Many think that the dolphin's strong sense of social conscience suggests some higher – perhaps superhuman – form of intelligence. Science has yet to prove that dolphins have the ability to reason, but some researchers are so convinced that we can learn from the wisdom of these creatures that they have attempted to open up channels of communication with them.

Psychologist Dr Louis Herman from the University of Hawaii allegedly made a breakthrough during the 1950s. He taught a pair of bottlenose dolphins to respond to more than 30 words strung together in hundreds of different ways. He noted a curious fact – that no matter how he used the word 'in' during conversations with them they responded in the correct fashion.

NATURE'S MYSTERIES

To him, it seemed that the dolphins had acquired an understanding of complicated English grammar.

Another study was conducted by American neurophysiologist John Lilly. He claimed to have even taught a dolphin to talk back!

During experiments at Marine World in Redwood City, California in the late 1970s, Dr Lilly wired a pool – home to a pair of dolphins called Rosie and Joe – with underwater microphones. When the dolphins made noises, their voices were recorded. Then one day an incredible message was captured on tape.

According to Dr Lilly, one of the dolphins spoke to him. 'Throw me a ball,' it said in a strange Hungarian accent. Sceptics, however, maintained that the microphones captured the voice of a Hungarian-born researcher working with the team. Indeed, many scientists to this day dispute the dolphin's ability to produce human sounds.

Nevertheless, dolphins do speak their own language – some researchers call it Delphinese. When an unacquainted pair are placed in nearby tanks they 'talk' together in strange high-pitched clicks, barks, rasps, squeaks and groans. Their way of conversing also seems almost human. While

DOLPHIN TESTS

That dolphins can be trained to perform tasks for man was proved by a strange US Navy experiment.

Films produced by a dolphin training programme in the 1960s show the playful creatures delivering messages to underwater laboratories at a verbal command from their trainer. Apparently, dolphins were also trained to act as 'frogmen', carrying packages which they later deposited on ships.

One man who was not happy about these revelations was Dr John Lilly, who voiced fears that these packages contained explosives or tracking devices. He was also outraged at the exploitation of the friendly animals.

To this day, the doctor fears that dolphins may be pressed into the service of the military – something that runs counter to everything we know about the peace-loving species. Lilly worries that such exercises may seriously impair mankind's future relations with dolphins should the communication channels ever be fully opened.

Experiments test the dolphin's abilities at speed and in an underwater language laboratory before each of the participants is given a tasty reward.

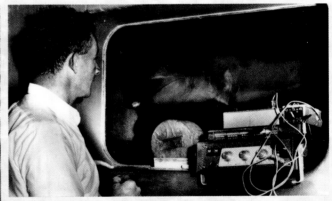

THE UNEXPLAINED

Dr John Lilly is willing to learn from an underwater culture.

one is 'speaking' the other remains politely silent.

Lilly, however, remains adamant that his discovery is no fraud. A man with Dr Doolittle aspirations, since the mid-1950s he has campaigned to prove that our own guidelines are wrong when it comes to judging dolphins' apparent genius and scorns those who refuse to contemplate the idea of dolphin intelligence.

Brave New World

In his 1978 autobiography, *The Scientist*, Lilly muses that man has much to learn from the dolphin 'life form' which he believes has developed its own culture and ethics in a separate world beneath the sea. Learning to speak the language of dolphins would constitute a first attempt to communicate with what he regards to be a race of 'alien' intelligent beings. In his view, this experience could be usefully applied if we ever happened to be approached by visitors from other galaxies.

The doctor claims, too, that knowledge of the dolphin way of life could also prove invaluable to man's own attempts to explore new solar systems. After all, the creatures' experience of weightlessness, learned from their life in the deep, gives them a natural understanding of how anti-gravity operates in outer space.

Significantly, Lilly argues that time is running out – that if man wants to learn from the dolphins, he will have to learn their language quickly, otherwise the creatures may discover that we regard them as just another of Earth's resources, placed here for our own exploitation. For the moment at least, there is a very real chance that dolphins will not like what they hear.

ALIEN ENCOUNTER?

The contact between dolphins and intelligent life forms from worlds beyond our own is a familiar science fiction storyline. But could this premise be based on truth?

In his book on dolphins, extraterrestrials and angels, American writer Timothy Wyllie claims that it is more than mere chance that so many flying saucers are seen entering and leaving the world's oceans. Wyllie says that his information comes straight – so to speak – from the dolphin's mouth. Apparently while visiting Florida in the late 1970s, he fell into 'telepathic rapport' with a group of the creatures.

The author believes the dolphins 'tuned into' his thoughts while he was musing over the UFO theory late one night. Seconds later, he saw a flying saucer appear on the horizon and watched a dolphin leap out of the water. In his opinion, this and other coincidences were orchestrated by the dolphins to prove their superior intelligence to him. Wild as the theory sounds, it certainly provides food for thought!

NATURE'S MYSTERIES

NATURE'S SECRETS

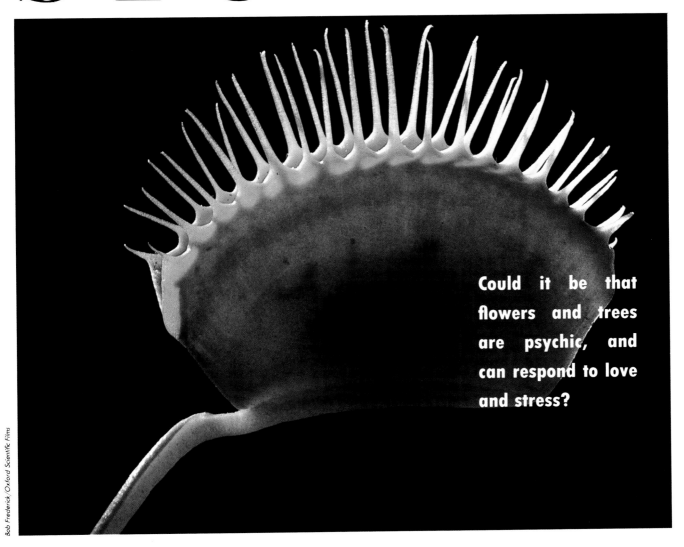

Could it be that flowers and trees are psychic, and can respond to love and stress?

Some people undoubtedly seem to have a facility for producing wonderfully healthy plants; others, meanwhile, claim that everything they try to grow withers, even when in suitable soil and watered according to the book.

Vibrations and Impulses
In the attempt to find out why there should be such major differences, researchers have produced a fascinating theory. It could be, they say, that some of us actually give out vibrations to which flora will respond in a particularly favourable way. And there is evidence to suggest that plants may not only react to external stimuli but will also transmit impulses.

Plant Sensitivity
It certainly seems to be the case that a person's state of mind can have a remarkable effect upon vegetation. Plants watered by hospital patients suffering from severe depression have been shown not to thrive nearly so well as when they were tended by someone in good health. Studies have also shown that if some of a batch of seeds are stared at in an outlandishly aggressive way over quite a long period, they will suffer inhibited growth whereas the control group develops normally.

Some experiments have shown that plants respond to their owners' varying emotions. Electronics engineer Pierre Paul Sauvin found that plants he had

THE UNEXPLAINED

Some gardeners firmly believe that plants, like people, thrive on encouragement and attention.

himself grown from seed and tended personally, would show a response whenever anything dramatic happened to him.

Perhaps, then, there really is sense in talking to one's plants to help them thrive.

Talking to Plants

Flatter your bulbs with kind words; give your plants pep talks; meditate over seeds before you grow them; speak kindly to their foliage and you will be amply rewarded, researchers have found. One American scientist even reported that he had succeeded in making an air plant that had never previously blossomed produce flowers by speaking to it lovingly. But such verbal encouragement is said only to work if you truly have faith in the results it will bring.

Sound and Growth

Music can also have a marked effect on growth. But plants are quite conservative in the main. Geraniums, for example, seem to thrive on Bach's Brandenburg Concertos; Brahms provides very positive vibrations, some claim; and contemporary ballads apparently also accelerate growth. Heavy Metal music, on the other hand, reputedly makes plants wilt. One Indian researcher found that if she performed a traditional dance for 15 minutes each day in front of young marigold plants, they would grow to half their normal size again; and a Canadian farmer had a particularly abundant harvest of soya beans after he had played soothing, melodious music to his crops.

Plant Intelligence

Experiments carried out under the auspices of the Soviet Academy of Sciences and at the University of Clermont in France seem to indicate that plants could well have memories of some kind, too. In one set of tests, one of a pair of cotyledons (seed leaves) was intentionally damaged with a needle. Both leaves were then removed from the plant to ensure that any ensuing reaction would not be due solely to the original pricking injury. Intriguingly, leaves on the side of the plant that had not been pricked with the needle fared far better.

Scientist and writer Dr. Lyall Watson has described how trees, such as willows, will react by producing certain chemicals when attacked by caterpillars or worms. Trees not yet infested will sometimes do the same – the conclusion being that there must be some form of warning system operating between them.

Professor Wouter van Hoven, a physiologist at South Africa's

To ensure a good crop of tomato plants it may be advisable not to mention eating them until after they have been picked.

NATURE'S MYSTERIES

Pretoria University, has also shown that the leaves of certain trees on which African antelopes commonly feed will produce a compound of tannins which is extremely bitter and which keeps the animals' appetites within bounds. The antelopes are thus often seen partaking of several bushes because the plants put up these chemical defences after only a few minutes when attacked for food in this way, and the animals move on in order to avoid the unpleasant tang as soon as it occurs.

Psychic Plants

There is also a certain amount of evidence to show that plants may sometimes be in communication with the spirit world. When Dr. Nandor Fodor, a psychoanalyst and psychical researcher, died in 1964, his wife soon began to notice that objects had started to move around their home, an occurrence she interpreted as an attempt by her late husband to reassure her of his spiritual survival. In addition, the roses they grew suddenly began to bloom for weeks on end, instead of

The Venus flytrap is a plant with the ability to trap insects. Its hinged leaves snap close when the surface hairs are touched.

PLANT AND TREE MAGIC

Many ancient magical practices involved the use of flowers or trees. Some were carried out to ensure good health: a horse chestnut, for example, was believed to protect against rheumatism if you kept it in your pocket; and warts could supposedly be cured by rubbing them with a piece of bacon which then had to be inserted in the bark of an ash tree.

Other plants were used to bring about good fortune. If you placed pure white flowers in a clear crystal vase and stood them on a window-sill on the eve of the New Moon, it was said you were assured prosperity for the whole of the coming month. By planting lilac, honeysuckle and almond trees in your garden, it was thought you could provide financial stability for the family. Silver birch, maple, holly bush or ash are still all thought to bring luck to a household, but many people remain convinced that monkey puzzle and plane trees are generally best avoided.

The elder tree has long been held to protect humans from the spirit world. The baptized were even said to become party to the secrets of witches if they bathed their eyes in the juice of its bark; and such were said to be the magical properties of the oak that, if you carried an acorn with you, eternal youth would be your reward. It was also thought wise to leave an acorn on a window-sill since it was said to have the power to ward off storms.

The Druids thought that mistletoe had special powers to ward off evil spirits.

Mistletoe, sacred to the ancient Druids, was also thought to have special powers, especially if found growing as a parasite on an oak. According to pagan belief, it had to be cut down with a gold knife on the sixth day of the New Moon and caught in a white cloth. It should not be left to fall to the ground or it would lose its effect. After cutting, the mistletoe would be placed in water, and the liquid later used as a charm to ward off evil spirits.

Fairy rings – bare patches of soil – were thought to result from fairies dancing round in a circle and said to be lucky, though you were advised not to step inside lest the fairies took revenge.

merely days, never losing any of their petals. Similarly, author Taylor Caldwell found that garden lilies that had failed to bloom suddenly came into flower on the very day of her husband's funeral.

The Venus flytrap plant snaps shut when its hairs are touched; and if it is wired up, electrical impulses are recorded whenever this occurs. Mimosa plants are also known to react quite dramatically to physical contact. Talk

THE UNEXPLAINED

sweetly to a flowering plant while giving out positive thoughts and, the green-fingered say, its scent will be all the more fragrant and its blooms more spectacular. Orthodox scientists and sceptics may sometimes scoff; but research does seem to indicate that communication in the plant world is more than a mere possibility.

When seedlings respond well to the tender loving care lavished on them, the grass in the garden is lush and green, the window box is a riot of colour and pot plants are magnificent specimens with glossy leaves, the gardener deserves to be called 'green-fingered'.

Certain trees and bushes develop a bitter taste when attacked or eaten. This restricts the feeding habits of the African antelope.

Roses that bloom in the freezing temperatures of winter are one of the miracles of Findhorn.

FINDHORN

Giant cabbages weighing as much as 19kg (42lb); delphiniums up to 2.5 metres (eight feet) tall; as many as 50 different types of herb and more than 20 kinds of fruit; roses that bloom in freezing temperatures; riotous beds of fuchsias, petunias, poppies and columbine: this is the horticultural miracle which Peter and Eileen Caddy lovingly created in the unlikely situation of a rubbish dump and caravan site at Findhorn in northern Scotland in 1962.

Its founders knew nothing at all about gardening when first inspired to come to Findhorn. The soil there is sandy and in theory should yield no more than gorse and a few grasses. The climate is very harsh (Findhorn is further north than Moscow), yet the community that settled at Findhorn has become something of a legend in its own right. They have broken all the so-called rules of nature by growing plants that orthodox scientists say should never stand a chance of flourishing in such poor earth and inclement weather.

The thriving garden is said to have been created by divine guidance. Dorothy Maclean, an original community member, for instance, made regular contact with the nature spirits or *devas* (a word derived from Hindu and meaning 'beings of light'). Her first contact was with a pea *deva* who taught her everything she would need to know about preparing the soil for an abundant crop. So, too, did the angels or *devas* of all the different plants at Findhorn.

Plants thrived there as if by magic. Visitors to the garden have reported that the plants seem to be fed by the very spirituality of the community. Said to be sited at a point on which a great deal of energy is centred, Findhorn remains a flourishing place. But its flora is not now quite as astonishing in size as it was in the past. Findhorn has become a community that explores a new way of relating to the whole planet through educational programmes and projects.

NATURE'S MYSTERIES

STEADY AS A ROCK?

Some stones can be slippery customers that stray far from home – others fall like rain or dance the night away

California's Death Valley boasts many natural wonders – including Racetrack Playa, a dry lake bed located almost 300 metres (1,000 feet) below sea level. Its parched mud surface differs little from others in the area, yet it attracts an increasing number of tourists each year. They come to see a 'tribe' of wandering stones, which appear to be able to move around of their own accord!

The rocks vary from small pebbles to irregular half-ton boulders. Experts have noted their ability to slide unassisted across the surface of the Playa. And when they 'walk', they flatten a discernible trail in the muddy crusts as they go.

No one has actually seen the rocks move, but their journeys have been monitored since 1968 by Dr Robert Sharp, a geology professor at the California Institute of Technology. By tagging the stones' positions, he has recorded that as many as 28 out of 30 rocks move. Incredibly, one travelled 262 metres (860 feet) in less than a year.

The phenomenon has excited much speculation, with theories ranging from the intervention of extra-terrestrials to the possibility that the stones have a will – or life – of their own. Dr Sharp, however, believes that the explanation is more down to earth! According to him, even a light shower of rain can give the baked mud surface a slippery coating. The powerful winds that often follow rainstorms in this region provide enough force to send the rocks skidding across the surface.

Rock of Ages
Wind and rain may – or may not – move the enigmatic Playa rocks, but the mysterious behaviour of stones and boulders has stirred mankind's imagination for centuries. Myths about stones which have strange powers are found in almost every culture. Some are said to heal, predict the future or even 'give birth'.

British folklore is rich with legends attached to both natural rock formations and mysterious stone circles that dance the night away. One colourful story relates to the Rollright Stones, a hauntingly beautiful Bronze Age 'cathedral' on the borders of Oxfordshire and Warwickshire. These megaliths are said to be the remnants of a noble army. Marching across England, the soldiers reached the village of Little Rollright where they were halted by a witch. She cried out to their king, 'Seven long strides thou shalt take! If Long Compton thou canst see, King of England thou shalt be!'

Confident that the village was just over the brow of the next hill, the king strode on ahead but found his view obscured by a hillock. With that, the witch turned him and his army to stone.

Exit stage left: one of the Playa rocks seeks a new path.

The story does not end there. These same stones are also said to take themselves off to drink from a nearby stream once a year. Even more sinister is the tale of the Lord of the Manor of Rollright, who decided to use the King's Stone to build a nearby bridge.

Moving it, however, proved difficult – a team of horses had trouble dragging it away. Soon after, the manor was plagued by ominous sounds and bizarre disturbances. The stone was blamed and returned to its original position. This time, a single horse dragged it easily – uphill!

Singing Stones
Today, such tales are dismissed as folklore. But researchers have uncovered some peculiar facts about the Rollright Stones.

Between 1978 and 1982, scientists working on the Dragon Project, a study of mysterious earth energies, found that certain stones in the Rollright group emitted strange ultrasonic pulses

---

THE WITCH OF SCRAPFAGGOT GREEN

Another British stone which lived up to its legendary reputation was that which stood at Great Leighs in Essex. The story went that the large boulder, positioned near a local road, pinned down the spirit of an ancient inhabitant – the witch of Scrapfaggot Green.

The US army used a bulldozer to widen the road in 1944, moving the boulder in the process. For three days, the village was plagued by poltergeist activity. Church bells rang of their own accord, tools moved themselves mysteriously and showers of stones hailed down on the village.

Eventually, ghost-hunter Harry Price was called in to investigate. He advised the villagers to return the boulder to its original spot in a midnight ceremony. When this was accomplished, all the disturbances immediately ceased!

NATURE'S MYSTERIES

◄ *The King's Stone of the Rollright group (below) gave a 'rock concert' for Dragon Project scientists (inset).*

at dawn. The King's Stone in particular seemed to be invested with the same 'talents' as the Memnon Colossus in Egypt, which 'sings' in a low-pitched hum at sunrise.

During other tests with a Geiger counter, the King's Stone gave off a positive reading. Even more startling were photographs taken with infra-red film of the rock at dawn. They revealed that the stone was glowing – surrounded by a rather faint but distinguishable halo. Investigations proved that the phenomenon was not caused by the sunrise, a fault in the film itself or the way it had been processed.

More research is needed before scientists can confidently say what causes these bizarre effects. However, some maintain that certain stones store energy like a giant battery and then release it when the conditions are right.

Hard-luck Rocks

Other mysterious tales of rocks appear to defy rational explanation. In his book on unexplained phenomena, *The Nature of Things*, Lyall Watson tells how an American airline executive Ralph Loffert and his family – taking a holiday in Hawaii in 1977 –

THE UNEXPLAINED

collected some stones from the slopes of the volcano Mauna Loa.

When they returned to America, a series of accidents befell all four children. Mark, aged 14, sprained an ankle, tore a cartilage, then broke an elbow; Danny, 12, fractured a hand then tore an eye muscle; Todd, 11, shattered an arm, developed appendicitis, broke a wrist and dislocated an elbow; while young Rebecca, 7, broke two teeth in a fall – twice!

Then the children's mother, Dianne, remembered that an old man had warned the family not to take any stones away from the island. She promptly bundled up the souvenirs and posted them back to Hawaii. But the eldest son continued to have inexplicable accidents. He had kept three stones. When they were returned, all the trouble ceased.

Further inquiries proved that such events were not uncommon: the National Park Service in Hawaii receives an average of 40 packages of the hard-luck rocks every day!

HAILSTONES FROM THE SKY

The question of whether stones have a life of their own remains open to debate, but some can certainly fly through the air. For centuries, falls of rocks and pebbles have been observed and recorded. Even today, odd showers of them plague people throughout the world.

In one case, five home owners who lived in a quiet suburban street in Birmingham were besieged by showers of rocks for more than three months in 1981. The backs of the adjacent houses in Thornton Road had to be boarded up or guarded with chicken wire to prevent the windows from being smashed.

Fortunately, no one was seriously injured. Local police kept watch for hidden attackers, using infra-red equipment and automatic cameras. They even tested the stones for finger prints but found nothing. The case was never solved.

A similar story was recorded a year later in Machakos, Kenya. Peter Kavoi and his family were eating dinner one night just before Christmas, when stones began to hail down on their roof. The same thing happened again the next day. The rocks appeared to materialize out of thin air before they rained to the ground. Sometimes they shattered when they hit the ground.

Local police were called in to observe the phenomenon which continued for the next six months. Prayers, exorcism and local witch doctors tried to stop the showers – without success. Once again, there appeared to be no explanation.

Chief inspector Len Thurley eyes the Birmingham rock shower (left inset) that damaged houses in Thornton Road. Chicken wire protects the windows (right inset).

NATURE'S MYSTERIES

CONSUMED BY FIRE

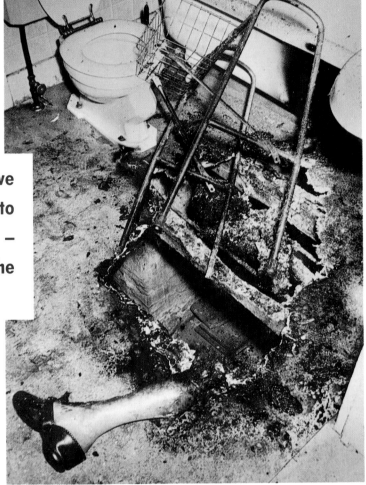

What causes live human beings to burn to a crisp – apparently from the inside out?

On the evening of November 12th 1974, Jack Angel – a travelling clothes salesman – parked his camper at the Ramada Inn in Savannah, Georgia, and settled down to sleep. He awoke later than usual – around midday – feeling rather odd. Looking down he was horrified to see that his right hand was burned black; other burns dotted his legs from groin to ankle.

In a state of shock, but – strangely – in no pain, Angel showered, dressed, and made his way to the hotel. There he discovered that the date was not as he expected it to be, November 13th, but the 16th. He had been asleep for four days.

Angel collapsed unconscious and was rushed to a nearby hospital where baffled doctors tried to make sense of what had happened. The skin on his hand was badly blistered, yet his clothes were undamaged. Further examination revealed the eeriest fact of all – Jack Angel had been burned from the inside out!

When the salesman eventually regained consciousness – by now he was in excruciating pain – he demanded an explanation for his condition. The perplexed doctors could offer none; staff at a nearby special burns unit were equally puzzled. His leg injuries healed, but his hand was so badly damaged that it had to be amputated.

Speculation mounted as to what had caused the horrific injuries. Jack Angel had not been smoking on the night of the fire,

121

THE UNEXPLAINED

Lucky to be alive – American salesman Jack Angel.

nor had he been scalded by boiling water which might have spurted from a kettle or the radiator of his camper. A firm of solicitors stepped in, believing they could sue the manufacturers of the camper for fire caused by an electrical malfunction. But a search for faulty wiring only proved that the electrical circuits were, in fact, perfectly sound.

The possibility of a freak lightning strike was also ruled out, since there had not been a storm on the night of the accident. So, after months of fruitless enquiry, only one conclusion could be drawn: Jack Angel had been a victim of the mysterious phenomenon known as Spontaneous Human Combustion.

Death by Fire

Authorities the world over have ridiculed Spontaneous Human Combustion – the sudden, virtual destruction of a human body by fire. But for hundreds of years there have been many reports of death by fire which completely defy scientific explanation.

One of the first recorded cases was that of Grace Pett, a fishwife from Ipswich, who was burned to death in her home in 1744. One April morning, Mrs Pett's daughter discovered that her mother had not been to bed. Downstairs, she found a body lying across the hearth 'appearing like a block of wood burning with a glowing fire without flames'. Neighbours dowsed the fire, but Grace Pett had been burned to ashes.

At first it was assumed that a stray spark from the chimney or a candle flame had ignited the woman's clothes while she dozed. However, the grate was empty and the candle had long burned down to a stub. The most bizarre element of Grace Pett's tragic demise was that the wooden floor beneath her body bore no sign of scorch marks. Being the 18th century, there were whispers of witchcraft and black magic, but the case was never solved.

A more recent instance involved Mary Reeser, a widow living in Florida. In July 1951, her body was discovered in her apartment – almost completely

THE DICKENS LINK

Victorian writer Charles Dickens was fascinated by Spontaneous Human Combustion. He studied known cases at length, and eventually used his knowledge in his novel *Bleak House*.

In chapter 32, Dickens describes the death of Krook, an evil drunk. His charred remains are discovered in his squalid room by Weevle and Guppy: '... there is a smouldering suffocating vapour in the room, and a dark greasy coating on the walls and ceiling... Here is a small burnt patch of flooring; here is the tinder from a little bundle of burnt paper... seeming to be steeped in something; and here is – is it the cinder of a small charred and broken log of wood sprinkled with white ashes... ? Call the death by any name... attribute it to whom you will... it is the same death eternally – inborn, inbred, engendered in the corrupted humours of the vicious body itself... – Spontaneous Combustion, and none other of the deaths that can be died.'

In describing the death of a villain, Charles Dickens leaves the reader in no doubt as to how it happened.

consumed by fire. All that remained of Mrs Reeser was her skull, which had shrunk to about the size of a cricket ball, her liver attached to the backbone, one foot still encased in a slipper and a layer of sticky grease. The rest of her body had been reduced to ashes, as had the chair she had been sitting on. The intense blaze had also melted her apartment's plastic door-handles; the room itself, however, was undamaged.

Dr Wilton Krogman, a forensic scientist who examined the body, was astounded. Intense heat usually causes a human skull to explode, but in this case it had shrunk. The apartment should have been burned severely, but only the chair in which Mary Reeser had been sitting was consumed. 'Were I living in the Middle Ages,' he remarked, 'I'd mutter something about black magic.'

Common Features

While no two cases of Spontaneous Human Combustion follow exactly the same pattern, many have common features. Several deaths have occurred in cars, including one strange case in the United States involving 27-year-old Billy Peterson from Detroit. He committed suicide back in 1959 by inducing carbon monoxide poisoning.

Nothing indicated that he had tried to set himself alight, yet his body was found horribly burned. The fire's heat had fused a plastic religious statue to the dashboard, but both the car and his clothes were otherwise undamaged.

That the victim's surroundings were left untouched is also a typical occurrence. Many have been found burned in bed lying on unscorched sheets or, in one case, in a hayloft surrounded by unburned bales. Few survive to tell of their experiences. Those who are found alive usually have no idea why the fire began and most of them die soon after.

The most unfathomable aspect of Spontaneous Human Combustion is the destruction of the victim's corpse. Because the human body is made up of more than 60 per cent water, it is extremely difficult to burn. Crematoriums pre-heat their furnaces to a temperature of 1,003 degrees Celsius, yet it takes between one and two hours to dispose of body tissue and major bones have to be ground to ashes. To reduce a human body to the liquid form it is often found in after Spontaneous Combustion

NATURE'S MYSTERIES

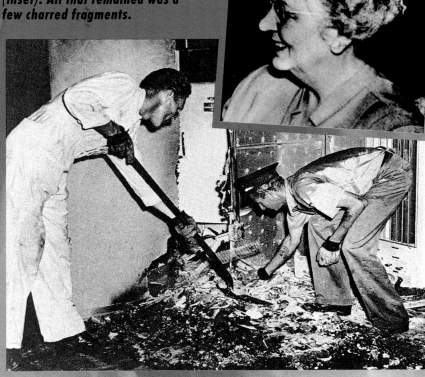

Clearing up after the sudden tragic demise of Mary Reeser (inset). All that remained was a few charred fragments.

Strangely, the surroundings of many victims of Spontaneous Human Combustion do not catch fire.

THE UNEXPLAINED

has occurred, it has been estimated that the temperature must exceed 1,648 degrees Celsius.

Loose Ends

Many theories have been put forward to explain the phenomenon but most of them only go half way towards providing an explanation. One early theory maintained that Spontaneous Human Combustion resulted from the ignition of poisonous gas within the body. However, to contain such gases, a body would have to be in an advanced state of decomposition. In all recorded cases, the victims were alive shortly before their ignition.

Other speculative research claims that a build-up of phos-

Experts continue to debate the cause of these bizarre deaths.

phagen – a naturally occurring substance similar to nitroglycerine found in muscle tissue – causes the body to become combustible. But no reason has been put forward as to how the fire actually starts.

Some physicists have theorized that ball lightning could emit micro waves similar to those produced in a microwave oven. Should this ever manifest itself near human beings, the result would be severe burning from the inside out.

Others, meanwhile, argue that the fires are connected with the earth's fluctuating geomagnetic field. Many reports of Spontaneous Human Combustion occur when the intensity of this field is at its highest peak.

It may well be, however, that not one but a set of conditions spark off such fires. Perhaps if coroners were not so keen to explain away bizarre fire deaths, more could be learned about this strange phenomenon.

The publishers thank Larry E. Arnold for original source material in the Angel case.

POLTERGEIST FIRES

Poltergeists have often been blamed for strange fires, and some people or places seem to be unusually susceptible.

In 1984, the offices of the magazine for the Institute of Health Assistance in Rome were plagued by a series of freak fires. The first victim was a 19-year old secretary, whose skirt suddenly and inexplicably caught fire while she was working quietly at her desk.

Soon after, there was another fire in the building's archives: 'It looked like a camp-fire,' said a clerk, 'but it started by itself. It burnt fiercely for a few minutes then it went out.'

There were two more fires in the same room that day, accompanied by ice-cold gusts of wind. The fire brigade could offer no satisfactory explanation. The staff blamed a 'ghost'.

The spook which started a fire in the archives – another unsolved mystery.

THE MIND

THE DIVIDED MIND

Why do some people feel as though they are possessed by the spirit of another person altogether?

Look in the mirror and who do you see? You see a face and an expression that you recognize as 'you'. But there is more to you than that. The way you behave, what you believe and your physical mannerisms are recognized by those you meet as your personality – an elusive thing that cannot easily be pinned down, yet undoubtedly the most important single thing about you. Without your personality, who would 'you' be?

Many people today believe that our personalities are simply the result of the way our brains work, our quirks and oddities being the sum total of all those strange electrical happenings in the brain cells. They also believe that if our personalities are really just a complex set of brain functions, when our brains die, then so do we.

Ancient people, however, believed that our bodies were simply hollows into which our personalites – or souls – were slotted. Some cultures saw this inserted

THE UNEXPLAINED

MEDIEVAL PERSONALITY TYPES

In medieval times, it was believed that there were four basic 'humours' in our bloodstreams that made us into one of four different types of person. They were the melancholic personality (earthy); the phlegmatic (watery); the sanguine (airy); and the choleric (fiery).

The four different types were believed to be instantly recognizable, although some 'mixes' were also thought to exist. The nearest we get to such a distinction today is in astrology, with its Earth, Air, Water and Fire signs, each of which is believed by some astrologers to be detectable from a person's face and mannerisms.

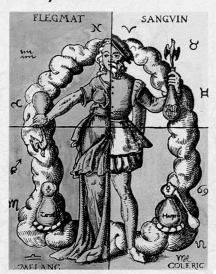

The four humours: melancholic, phlegmatic, sanguine and choleric, shown here with their corresponding zodiac symbols.

persona as a bird that flies away on bodily death; others, as a form of ghost or apparition that gradually disintegrates.

Split Personalities

But how do we explain it when something goes wrong? Most of us see personality as one, instantly recognizable character. Carol is always Carol, even though some days she is bad-tempered and on others she is as sweet as pie. But what if Carol suddenly announced that she was actually called Liz, spoke differently and claimed to have lived an entirely separate life from Carol?

Labelling such odd and disturbing behaviour as 'schizophrenia' actually does nothing to explain it, although at least it sounds more clinical than saying that Carol is 'possessed' by the spirit of someone called Liz. Yet there are many modern psychiatrists who are returning to the idea that the patient's claim of being possessed should be taken at face value. In their opinion, therapy then becomes far easier.

Phobias

Many phobias or unexplained and irrational fears have been shown to respond to 'past-life therapy'.

◀ *Personality change features in R. L. Stevenson's story 'Dr Jekyll and Mr Hyde'.*

Medusa, a famous figure of Greek mythology, had hair made of snakes, a real nightmare for those with a snake phobia.

THE MIND

MASKS

The human face often gives far too much away. So to hide revealing expressions, masks have been worn in ritual and ceremony the world over. A mask is a frozen expression – behind it, an individual may feel many emotions but only show the one that is carved on the mask.

Ancient priests and tribesmen wore masks to emulate the gods. Some were made of gold and inlaid with precious jewels; others were made from animal skins and painted with elaborate designs. They served the purpose of distancing the wearer from ordinary people, by making him appear strange and mysterious.

In witchcraft and magic, masks were worn to aid invocations of the gods. In medieval times, someone wearing a mask of a horned god might well have been responsible for tales of the Devil himself. The witches were sometimes drugged with hallucinogens, so the effect of the horned mask must have been horrifically magnified.

Wearing a mask means you can behave differently: inside it, you feel free from the necessity to conform. The famous 19th-century writers, the Brontë sisters, at times wore masks as children – their father's way of getting at what they really thought.

▲ *The tradition of wearing mysterious-looking masks continues today at the Venice Festival.*

◀ *A bronze ceremonial tribal mask, which would cover the entire head, from Benin, West Africa.*

During this, the patient is taken back through hypnosis to what seems to be a previous life or incarnation. Terrible things that happened in that life, it is believed, can carry over into the present, surfacing as phobias.

Reincarnation

Past-life therapy is often amazingly successful. One young woman who was terrified of flying apparently 'became', under hypnosis, a young pilot who had died in the Battle of Britain. It seemed that all his fears about that mission and his death had been unresolved and carried over as a phobia into her next life as a girl.

The idea that we are the sum total of all our previous lives is gaining ground. Today, more people than ever before in the West believe in reincarnation – up to 60 per cent in France and parts of North America, according to a survey made in the early 1980s. But ideas about reincarnation differ quite wildly.

Some think that we carry buried memories of all our lives – which may be hundreds, straddling the centuries – deep within

Schizophrenics with 'split personalities' often appear to have more than one identity.

the present personality. Odd incidents or meetings, they believe, can trigger off some of these memories; or we may experience them in dreams, or as flashes of intuition. Others think we are only the sum total of especially significant lives.

Matching Marks

In India, where the belief in reincarnation is particularly strong, many children have been born claiming to be the reincarnations of people who died shortly before their births. They sometimes even have scars or marks matching those of the dead person, which seems to confirm their beliefs.

Another common phenomenon is known as 'dissociation' – where parts of the personality act as if they belong to a separate person. In much the same way, spiritualist mediums of old, and contemporary Californian-style 'channellers', claim to be taken over by an assortment of invisible characters who pass on messages to the living audience.

THE UNEXPLAINED

It may sound alarming and even dangerous to have a personality that can become fragmented, possibly rather like having the whole cast of a soap opera in your head. But sometimes, as in the case of 'multiple personalities', there is thought to be a sound, almost canny reason for what appears to be very disturbing behaviour.

Victims of sexual abuse, for example, have been known to produce a character who is carefree and flippant, or strong and assertive – totally unlike the poor child herself who is terrified and desperate.

Multiple Personalities

The new personalities tend to stay only as long as the shock lasts, although they may come back from time to time, and are thought to be a dramatic way of surviving the trauma.

Up to 30 different 'personalities' have been recorded as 'living' in one person, each of them very different and most of them clamouring for attention at the same time. But in most cases of multiple personality, there are just one or two 'extras' that take over when the 'host' personality has been through a crisis.

Although many of the apparent mysteries of personality have been accounted for, there remain a number of strange phenomena that may never be fully understood.

FACE-READING

We often say that we take someone 'at face value', meaning that we accept the image they put forward and do not look below the surface. But the ancient art known as 'physiognomy' decrees that we can deduce a considerable amount from the details of the human face – the set of the ears and the tilt of the nose, for instance.

The true physiognomist draws up a chart of all a person's outward facial characteristics, crediting and debiting points according to a complex system. He will then arrive at his conclusions. While a certain type of jaw by itself might not point to an 'ardent' or 'degenerate' type, a particular length of head or ears with 'poor' or 'bold' rims would clinch the matter.

Some of these old theories carry on, perhaps because they are based on commonsense. For example, a furrowed brow in someone relatively young would indicate a natural worrier; and if it is found with a tight mouth that is turned down at the corners, then the chances are that the worrier is also spiteful and petulant.

◀ *Which personality traits do the physical features of these faces from a painting by Bosch suggest to you?*

▼ *In Taiwan, street fortune-tellers offer character readings and divine the future by interpreting your facial features.*

THE MIND

HIDDEN POWERS

Do all of us have the potential for extra-sensory perception – and if so, how can we learn to develop it?

Henry Dakin/Science Photo Library

Predicting the future, scrying and telepathic communication are just three of the mysterious arts which fall within the realms of Extra Sensory Perception. While many people believe that such abilities are somehow linked with the supernatural, most modern practitioners claim that ESP powers lie dormant in almost all of us – and that training is all it takes to bring them out.

Psychics argue that ESP is a primary sense like touch or taste, and bears no relation to intelligence or education. Evidenced in wild beasts which sense danger and changes in the weather, or in pets which travel long distances to find their owners, is it a gift which mankind has somehow forsaken along the road of evolution?.

The world's major powers are said to have spent vast sums of money researching ESP over the last 20 years. Clearly, a sixth sense could reap great rewards when applied to government strategy and dealings, but what about on an individual level? What methods can we use to develop our own hidden powers?

The decision to begin flexing

129

THE UNEXPLAINED

the psychic muscles is not one to be taken lightly, and new students are often warned that it can radically alter both the way they think and react to others. While an ability to predict the future can bring visions of good fortune, it may also hint at unpleasantness. With foresight, however, such problems can usually be avoided – a rewarding experience in itself for those harnessing their hidden powers.

The first stage of any psychic journey is to relax, as tension is said to block the natural flow of alpha waves – the electrical activity in the intuitive part of the brain associated with ESP. Quite simply, this means 'turning off' from the immediate surroundings. Techniques such as yoga, transcendental meditation or Silva Mind Control can all be helpful in achieving the altered – some would say higher – state of consciousness required for successful ESP experiments.

Telepathic Contact

Controlled scientific experiments have shown that mind messages can be sent over vast distances.

Such telepathy is thought to be easier between people who know each other well, as shown by the many cases of husbands and wives, parents and children, and brothers and sisters who claim to have 'contacted' each other prior to a disaster. Similarly, it pays to begin your own experiments with someone who is close to you, always remembering to respond instinctively rather than trying to 'think' telepathically as you would if solving a problem.

For the layman, Zener cards are perhaps the simplest way to test thought transference. Available commercially (though you can easily make them yourself), they come in packs of 25 with an equal number of identical circle, star, wave, square, and cross patterns. The cards are shuffled and viewed at a time by one person – the *transmitter* – who mentally pictures each image and attempts to relay it to their partner – the *receiver* – so that they can guess which card is being held. The probability factor for a successful guess is one in

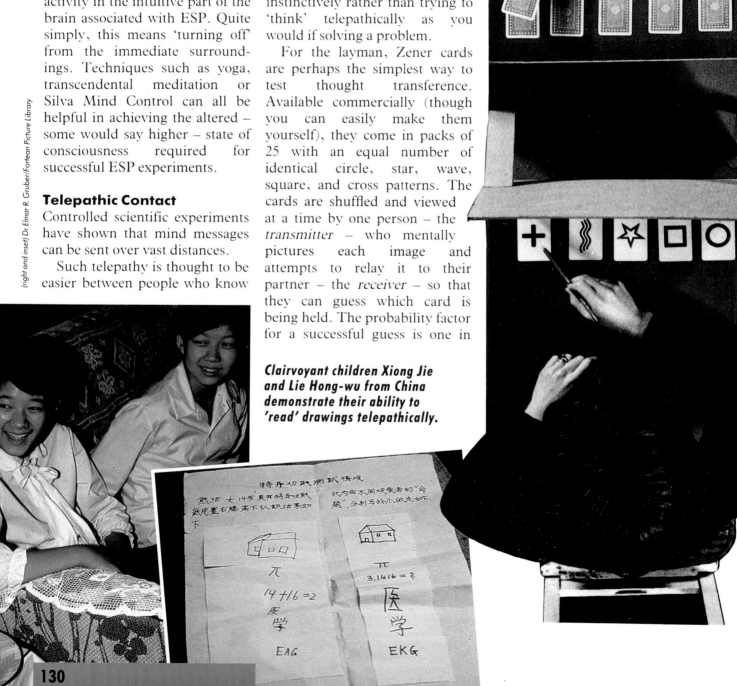

Using Zener cards can help two people develop an ability to communicate telepathically.

Clairvoyant children Xiong Jie and Lie Hong-wu from China demonstrate their ability to 'read' drawings telepathically.

THE MIND

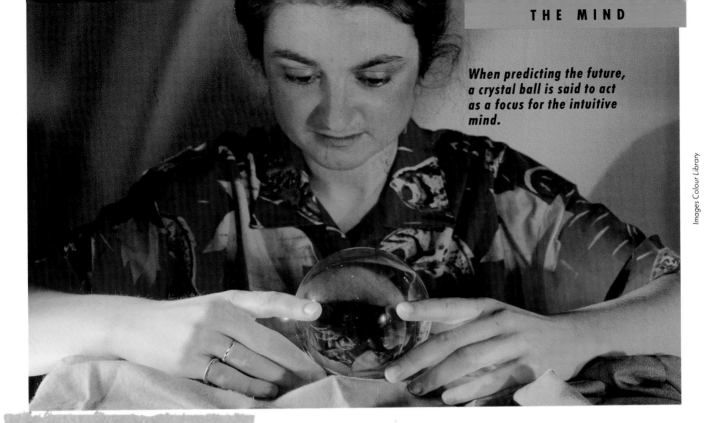

When predicting the future, a crystal ball is said to act as a focus for the intuitive mind.

THE PSYCHIC'S DIET

Food is thought to play an important part in developing the powers of ESP. Most psychics find it impossible to operate effectively when they feel tense, and since some nutrients affect our state of mind, they look to foods which contain nutrients that will lift depression or instil a sense of calm.

Despite little research on the subject to date, some psychics say that a strict vegetarian diet diminishes their healing powers while increasing those of telepathy. They suggest avoiding foods like cheese, eggs, chocolate, currants, nuts, bacon, lamb, coffee, chicken and grapefruit juice.

Instead, try including honey, skimmed milk, fish and shellfish, peaches, apples, carrots, pears, wholemeal bread, lettuce, lentils, beef, tea and garlic as part of a balanced diet. Regular exercise can also ease the tension that is said to impair our hidden sixth sense.

---

five; anything exceeding this scoring rate points to some kind of genuine telepathic ability.

Pictures or other images can also be used to prove telepathic links – do a series of drawings, then see if your partner can 'tune in' to your mind and describe the impressions. Try swapping roles, receiving messages as well as sending them.

Making Predictions

Thinking of a friend only to find that they phone you shortly afterwards, or dreaming of an event that subsequently happens both suggest an intuitive talent for predicting the future. Keeping a personal logbook of such occurrences and referring to it regularly will boost your confidence and spur you on to develop your powers further.

To experiment, try guessing what a friend will be wearing the next time you meet, or what the next tune will be on the radio. Tell someone about your prediction or write it down, paying no regard to early errors. Practice is the most important thing, and as your instinctive talent grows, so your success rate should improve.

The Crystal Ball

Many people still regard scrying – the art of using a crystal ball to see into the past and future – as a blatant form of witchcraft. According to current thinking, however, the ball is no more than a focus for the intuitive mind, where personal thoughts and visions are projected in an optical effect known as *blobs*. These can be cloudy mists, abstract symbols, or even real-life scenes.

While expert scryers do not worry about their surroundings, people new to the art usually find a quiet, darkened room less distracting. A dark cloth placed beneath the ball will also prevent unwanted reflections. Be patient though – it may take several sessions before you see anything.

Scrying traditions have been systemized for students who have initial difficulty finding any meaning in the mists. Green or blue cloudiness is said to herald joy; red, yellow or orange may signal disaster. A mist that rises answers a question with a yes; descending, it means no.

THE UNEXPLAINED

Psychometric Impressions

The art of picking up impressions about a person from something they own is a form of ESP known as psychometry. Practitioners believe it is possible to receive personal information from the energy stored in simple objects like earrings, watches, or any item which has been worn by the subject for several months. The technique has even been used by psychics to help police detect crimes and find missing persons.

To test your own powers of psychometry, hold a treasured possession belonging to a friend gently in the palm of your hand. Without thinking about it, try to remain conscious of the item and see what intuitive messages flood into the mind.

Unlocking the Door

The key to developing powers of ESP is to believe wholeheartedly in whatever you are trying to achieve. Scientific experiments have proved that those people who acknowledge the existence of their own psychic ability actually achieve better results in clairvoyance tests than those who profess to be sceptical about the entire subject.

G. Gladstone/Image Bank

COLOUR SENSATION

One of the strangest psychic abilities is demonstrated by people who can judge a colour without seeing it. But is it really so peculiar that some claim to 'feel' certain shades?

Science has shown that light waves reflected by colours each have a different rate of vibration. Red vibrates slowly; violet quickly. Colours also produce definite sensations: blue evokes coolness, space and calm; red creates warmth. Blind people are taught to 'see' colours by responding to each shade's vibratory rate. Meanwhile, psychics sense colour changes emanating from energies in the body, using them to identify spiritual or physical imbalances.

Experts claim the ability to feel a colour can be mastered with practice. Red is sensed as a burning sensation, while orange is less hot and yellow feels tepid. Violet is said to be cold and may even pinch. To test for yourself, wear a blindfold and ask a friend to present different coloured objects or cloth. Some people find that testing with their noses, earlobes or tongues produces more accurate results.

By ignoring things simply because they seem improbable, it seems we may curb the spontaneity which is so important in the world of the true psychic. As children, we are taught to think and operate solely on the basis of factual evidence and may even be scolded by our parents for 'imagining things'. Yet if that simple, child-like spark of awareness can be rekindled – and the mind trained to listen to its own instinctive inner voice – then perhaps one day all of us will be able to bring our own elusive 'sixth sense' into daily use.

TELEPATHY AND THE GANZFELD TECHNIQUE

During an opinion poll run recently in the United Kingdom, more than 70 per cent of those questioned said they believed that messages could be conveyed telepathically.

One experiment devised to test the phenomenon is known as the Ganzfeld ('whole field') technique. It was first used with spectacular results by Charles Honorton, at the Parapsychology and Psychophysics division of New York's Maimonides Medical Center, and his colleague Ellen Masser, who possessed known psychic abilities.

Honorton became the transmitter for the experiment and was shown projected pictures of a nightclub. Meanwhile, Masser relaxed in a room bathed in pink light, her eyes covered by opaque half ping-pong balls. Wearing earphones which produced a synthesized noise similar to running water, she soon began to describe images resembling those Honorton was trying to convey. The odds against such accuracy were about 100,000 million to one.

The Ganzfeld technique – scientific proof of the powers of ESP.

Dr Elmar R. Gruber/Fortean Picture Library

MAGIC MEDICINE

THE MIND

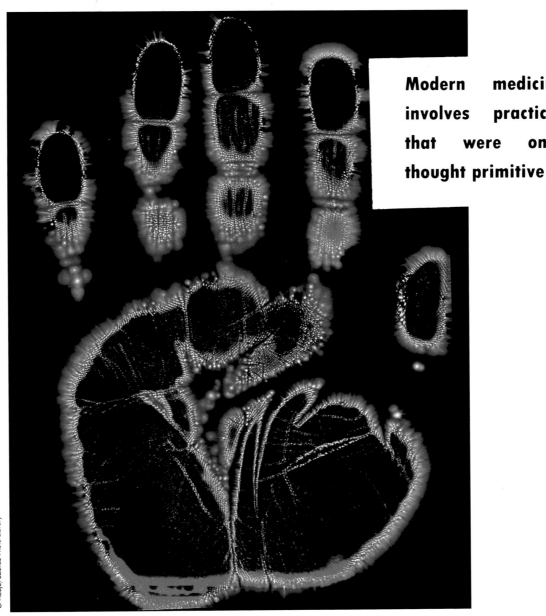

Modern medicine involves practices that were once thought primitive

Conventional Western medicine today is associated with increasingly complex surgery, equipment and drugs that would have seemed baffling to the doctor of a century ago. Yet despite such sophistication, more and more patients are beginning to realize that a strictly mechanical approach to the human body is not necessarily the answer.

Dissatisfaction with conventional methods has fuelled the recent boom in alternative medicine, which is ironically in many ways a re-evaluation of so-called 'primitive' attitudes to healing and treatment – attitudes which many thought obsolete in the face of scientific advance.

Today, holistic medicine – treatment of the person as a whole, body and mind – is an increasingly popular approach. But is this really so new, even if earlier attempts were somewhat more crude, involving attempts at magic? Is modern medicine reverting to ancient practices?

THE UNEXPLAINED

Surviving Egyptian documents dating from about 1500 BC describe the use of such practices in medicine. The doctor-magician would harangue the demon within the body of the patient and cast spells to induce it to leave. If unsuccessful, the unfortunate sufferer might be obliged to swallow a filthy concoction which was intended to make the body untenable for the spirit. This might consist of excreta, live or crushed insects, or other unpalatable forms of animal or vegetable life. A magically charged piece of cloth might also be attached to the patient's body.

▲ *The Egyptians had many practices to ensure health during life, and the soul's well-being after death.*

At the same time, the Egyptians were also making great advances in medical science. Surgical treatments for wounds of the head and chest had been formulated, and the treatment of intestinal worms and parasites, boils and cysts was practised as early as 1750 BC.

To ancient civilizations, and indeed as late as the 18th century, there was no anomaly in the combination of apparent 'mumbo-jumbo' and strict science. The inter-relationship of spirit and body was seen as a reality, and attention to the spirit was as important as care of the human body itself.

SHAMANISM AND HEALING

The figure of the shaman or witch doctor as the mediator between his people and the spirit world is universal both in myth and in so called 'primitive' societies. He is both spiritual counsellor and healer, and his task is one of total service to the community.

Health to the shaman is directly related to spiritual well-being, and so cure does not necessarily require just medicine for the physical body. The shaman, therefore, may diagnose an illness as being due to revenge on the part of an offended ancestor spirit who can only be placated by a purification rite or special ceremonial.

Shamanic knowledge of the human condition is gained through a long and painful initiation process whereby death and rebirth are

▲ *Equipment of the shaman includes a staff and crown.*

▶ *Small bone amulets are used in many curing rites.*

experienced on the psychic plane.

Such spirit journeys are usually taken with the aid of an hallucinogenic plant or trance dancing. The summoned power is then used to pull out sickness psychically.

Thus shamanism, perhaps the most ancient of healing practices, echoes our modern holistic approach in its regard for mind and body. Perhaps we are not so far removed from our Paleolithic ancestors after all.

THE MIND

This concept lives on in the renewed interest in pyschosomatic illnesses, of which asthma is one of many that are recognized to have their origin in the psyche.

Greek medicine owed much to the Egyptians. In 400 BC Hippocrates, dubbed the father of medicine, denounced magic, although some of the writings ascribed to him are devoted to the use of dreams as a diagnostic tool. They state that 'If the dreamer flies in fright from anything, this means an obstruction to the blood as a result of dehydration,' recommending that 'it is then wise to cool and moisten the body.'

The Greek philosopher-physician Empedocles formulated the concept of the four humours – black bile, yellow bile, blood and phlegm – each of which combined one of the four elements (Fire, Earth, Air and Water) and one of the four qualities (dry, wet, hot or cold). It was believed to be an imbalance in the humours that created illness. Thus too much 'blood' in the system was cured by bloodletting, using medicinal leeches.

KIRLIAN PHOTOGRAPHY

In 1939, a Russian, Semyon Kirlian, discovered by accident that, if an object is placed on a photographic plate and subjected to a high-energy electric field, an image will then result.

Such photographic images, with their characteristic 'halo' of discharges, were immediately thought to have performed the seemingly impossible; for they captured visually the 'invisible' aura surrounding all living matter as widely described by mystics down the ages.

The scientific establishment seems unconvinced, however, claiming that a non-physical manifestation can hardly be picked up by material instrumentation. But subsequent development of the process by the inventor and other scientists has proved that, even if Kirlian photography does not depict the aura itself, it seems to show up changes in electromagnetic fields, which enables it to be used as a diagnostic tool.

Recently, a Romanian doctor has even refined the method to detect cancer at an earlier stage than conventional x-rays and scanners. It seems certain that Kirlian's method of photographing the aura will continue to arouse controversy.

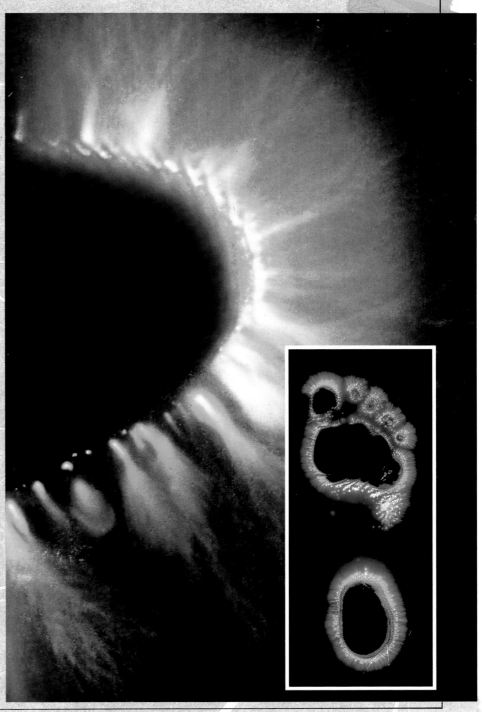

No external light source is used in these photographs showing the 'corona' emitted from a fingertip and the sole of a foot.

THE UNEXPLAINED

The Renaissance revival of learning was responsible for re-introducing classical medical practice to the West which was to endure for several hundred years more.

Paranormal Powers

Perhaps the most remarkable healer of all time was Edgar Cayce. Born in 1877 and a farmer's son from Hopkinsville, Kentucky, he showed paranormal powers of perception at an early age. By his twenties, the extraordinary talent which he had for diagnosing with only the patient's name and address for guidance was widely recognized.

Until his death in 1977, thousands benefited from the treatments prescribed in his trance readings, many of which on superficial evaluation seem extraordinary. The doctors who attended a seriously burnt two-year old, for instance, were told to apply poultices of tannic acid. The girl was not only cured but barely scarred.

Today we are seeing the rapid growth of a quite revolutionary approach to healing, whereby therapies formerly derided as the exclusive domain of cranks – such as homeopathy, acupuncture and colour healing – are attracting worldwide attention. The traditional view that illness is a physical discomfort to be eradicated as soon as possible is gradually changing, as a greater appreciation of emotional causative factors come to the fore. Healing and spirituality are joining forces once more.

FAITH HEALING

Healing, achieved by prayer or the laying on of hands without recourse to conventional medicine, has been practised for centuries.

Christianity does not have a monopoly on faith healing, however. The ancient custom of marking a prayer by tying pieces of cloth around curative wells and bathing in the waters predates Christianity and is to be found from Mexico to India.

Cure through the laying on of hands, as practised by modern healers, was also a power attributed to kings. Until the reign of George I, the British monarchy was traditionally supposed to have the power to heal *scrofula* or the King's Evil, a hideous skin disease. Charles I apparently cured a hundred sufferers at Holyrood on St John's Day in 1633.

Although the investigation of faith healing is a complex and often controversial issue, it seems that serious attempts to heal in this way will probably survive as long as hope itself.

At Lourdes, the famous Roman Catholic healing shrine, thousands of 'miraculous' cures occur.

TWIN DESTINIES

Some twins seem to live as though they are one. Is it mere coincidence or an uncanny link?

THE MIND

Where twins are concerned, the hand of fate appears to work in many mysterious ways. Not only are some difficult to tell apart, but many share an uncanny bond which can at times seem telepathic. This is not just confined to identical twins; non-identical (fraternal) pairs can also experience the same bond but usually to a lesser extent.

Alike As Two Peas
Identical twins Greta and Freda Chaplin, for instance, are alike in virtually every way. These sisters from York hit the headlines when they were charged with breaching the peace in 1980 by harassing a neighbour. But it was not this that shocked followers of their court case. To everyone's amazement, Greta and Freda appeared to speak in perfect unison!

Although tests proved that, in fact, one twin talked a split second before the other, people continued to puzzle over their mysterious empathy. Greta and Freda thought and spoke as one, and ate their meals together, simultaneously lifting forks full of the same food to their mouths. Both wore identical clothes, and if they received gifts of socks or gloves in different colours, they wore one of each. When one received medication, the other demanded to be given exactly the same prescription.

'We are so close that we are really one person,' the pair said when interviewed by a magazine journalist. 'We know exactly what each other is thinking because we are just one brain.'

Apparently, Greta and Freda

THE UNEXPLAINED

Chaplin's mirror-image lifestyle was encouraged from an early age by their parents. But bonds between twins can exist even when they have not spent all their lives together.

Parallel Lives

Jacqueline and Sheila Lewis are just such a pair. Adopted by different families at birth, neither knew that the other existed until they met by chance. The twins were reunited in a Bristol hospital where they were being treated for an identical, rare skin disease in June 1976. Through sheer coincidence they were placed in the same ward. Then the fun began.

Nurses were unable to tell the pair apart and checked their medical records to find an answer. There, they learned, that both women were born on the same day and had similar early backgrounds.

Jacqueline and Sheila soon discovered that they were identical twins. They both had double-jointed little fingers, moles on their knees and marks on their necks. Both had suffered from kidney trouble and had been bothered by pains in their legs for some time. Incredibly, Sheila's husband had died two years earlier – on the very same day that Jacqueline was divorced.

A number of other pairs that have grown up apart have learned later that they have lived 'parallel' lives. Researchers at the University of Minnesota studied 71 pairs of identical and fraternal twins who were separated at birth. Many of the identical twins could boast amazingly similar experiences.

Some scientists believe that similar medical histories and personalities can be explained by genetics. They have discovered that most identical twins teethe at the same time and go bald together. Girl twins often begin to menstruate on the same day.

Two of a kind: identical twins Greta and Freda Chaplin.

Unusually Similar

Genetics alone, however, can not explain all the unusual coincidences discovered by Bridget Harrison of Leicester and Dorothy Lowe of Burnley, Lancashire – identical twins who were separated shortly after birth. Reunited in 1979, after 34 years apart, each was wearing seven rings, while they also had two bracelets on one wrist and a watch and bracelet on the other.

The pair discovered that both had stopped taking piano lessons at the same age, each had owned a cat called Tiger and that one had named her son Richard Andrew, the other Andrew Richard. Bridget and Dorothy had also kept a diary for one year – 1960 – and chosen a book of the same type and colour to write in. They even left the same days blank!

More than 3,500 pairs gather at Twinsburg, Ohio for an annual twins' convention.

THE MIND

So alike are Bridget Harrison and Dorothy Lowe that they could almost be the same person!

More bizarre are tales of twins who share each other's pain. One fascinating case involved Silvia and Marta Landa who came to the attention of the Parapsychology Society of Spain at the tender age of five. Silvia had burnt her right hand on a hot iron and Marta felt the pain 12 miles away. Both later developed similar burn scars on their right hands.

This phenomenon has also been found in fraternal twins of a different sex. Each time Yvonne Green, for example, gave birth, her twin brother Christopher Gool 'felt' her labour pains 300 miles away. Another time, when Christopher hurt his arm in a brawl, Yvonne fell over and had to go to hospital for treatment.

Twin bonds as strong as these can, unfortunately, even prove fatal in a few cases.

Australian Joyce Crominski had a tragic tale to tell of the death of her identical 19-year-old twin sisters, Peg and Helen. Peg died on the way to hospital after being involved in a head-on car crash and fatally injured when the steering wheel pierced her chest. Asleep at home, Helen awoke at the same moment as her sister's accident with such terrible chest pains that her parents called an ambulance. She, too, died later.

'Their thinking was so alike that they would buy each other the same books and dresses for their birthday,' Mrs Crominski told *Truth* magazine in 1979.

Peg and Helen's mental empathy had been the subject of research at the University of

ANOTHER LIFE?

Twins Gillian and Jennifer Pollock were the focus of a tale which caused much controversy, based as it was on the personal testimony of their father, John. He claimed they were the reincarnations of their twin sisters who were killed by a car in May 1957.

After the 11-year-old twins, Joanna and Jacqueline, were hit by a motorist in Hexham, Northumberland, John Pollock said he prayed for proof of rebirth. Within months, his wife was pregnant and again gave birth to twins.

Jennifer, the younger twin, was born with a mark on her forehead that resembled a scar which Jacqueline had received as a two-year-old. She also had a birthmark on her hip which was identical to one that Jacqueline had possessed.

Various bizarre events happened. Jennifer and Gillian moved from Hexham with their parents when they were only four months old. Yet when their mother took them back for a visit three years later, they apparently recognised the family home and their dead sisters' school.

About a year on, their father happened to show them two dolls which had belonged to their dead sisters. Jennifer said: 'Oh that is Mary and this is my Susan. I haven't seen her for a long time.' The names were those given to the dolls by the dead twins.

The last inexplicable incident happened when Gillian and Jennifer were playing in the garden one day. A car engine started and they both began to scream. John said he found them crying: 'The car, the car, it is coming at us.' It was pointing at the exact angle of the vehicle that had hit Joanna and Jacqueline.

Were Gillian and Jennifer Pollock reincarnations of their sisters?

139

THE UNEXPLAINED

Dunedin in New Zealand for some time before their deaths. They could apparently transmit symbols on cards and pictures to each other even when sitting in different rooms. The scientist who studied them, said Mrs Crominski, believed Helen died when Peg's pain and shock after the crash was 'transmitted' to her.

Intuitive Bond

Telepathic links between twins, however, are extremely difficult to prove. A recent one-year study on Twins and ESP at Bristol University revealed scant proof of the paranormal, but it did show that twins had a remarkable ability to think alike.

Twins scored no better in controlled ESP tests with Zener cards than ordinary brothers and sisters. But when they were allowed to choose their own picture or image to 'transmit' to their twin they scored higher.

However, laboratory conditions do not resemble the highly-charged emotional moments when many twins claim to share their peculiar bond. Although conclusive evidence is as yet lacking, there seem to be too many links between pairs to put it all down to coincidence. Perhaps some really are born to share the same destiny.

Double take: two sets of identical twins wed.

MYTHOLOGICAL TWINS

Twins have always been thought of as mysterious and for this reason have figured extensively in mythology, folklore and religion. Stories about them can be found in most parts of the world and many centre around tales of two brothers who in some way help mankind. India's twin Hindu gods, for instance, are called the Asvins and they are associated with fertility, rain-making and healing. The legends of ancient Greece and Rome also had their own twin gods – Castor and Pollux, sons of Jupiter and Leda, who on their death became elevated to twin stars in the constellation of Gemini.

For sailors, a glimpse of Gemini during a storm was supposed to be incredibly lucky. The stars Castor and Pollux were thought to appear as lightning around the mast of a ship – an electrical phenomenon known as Saint Elmo's fire. As 19th-century writer Macaulay put it:

*'Safe comes the ship to harbour
Through billows and through gales,
If once the Great Twin Brethren
Sit shining on her sails.'*

Ancient statues of Castor and Pollux in Rome. Above them, the Gemini constellation, supposed to be a lucky omen for sailors if they caught sight of it during a storm.

THE MIND

CHILDREN OF ANOTHER WORLD?

Do some children exhibit powers that all of us possess, yet suppress or push into the background when we become adults?

When parapsychologist Ernesto Spinelli tested the ESP powers of children, he found that the younger they were, the higher they scored. He concluded that all children have varying degrees of psychic ability, but that this power becomes submerged as they grow older.

In most cases the submerging process happens naturally as the pressures of life take over. Sometimes it is not only school and social commitments that seal off the mystic channel, but the active discouragement of parents. 'Of course, there's no one here,' alarmed parents might tell a child who they catch talking to an 'imaginary' friend.

Childhood Visions

William Blake (who later wrote prophetically, see page 85) was one child whose supernatural awareness was too strong to suppress. He had his first vision at four – when he saw God put his head through the window – and they became a regular occurrence.

Returning from a walk, Blake told his mother he had seen the prophet Ezekiel sitting under a tree. Angry at what she thought were lies, she gave the child a beating, but punishment was not enough to stop Blake's visions. When he was nine he saw 'a tree filled with angels', while as an adult he wrote of attending a fairy's funeral and of beholding heaven's glory in a wild flower. The experiences stayed with him for the rest of his life.

THE UNEXPLAINED

Angry Spirits

One paranormal phenomenon which is almost exclusively centred around pre-teenage children is poltergeist activity, when unseen destructive entities terrorize households by moving furniture, levitating objects, rapping walls and committing other malicious pranks. Spiritualists believe the children act as 'mediums' for these invisible rogues. However, many psychic researchers hold that it is body energy radiated prior to the onset of puberty – a highly emotional period for any child – which lies at the root of the mystery.

A classic case involving an 11-year-old Scottish girl named Virginia Campbell started when she was kept awake by a noise like a rubber ball bouncing around her room. The next day, while she was having tea with relatives, a heavy sideboard lifted a couple of inches and apparently 'jumped' into the room. That night Virginia's bed started vibrating violently, loud banging emanated from the headboard, and a vicar who had been called to the scene testified to seeing a linen chest move jerkily across the floor.

The poltergeist activity even followed Virginia to school – and later to a relative's house where she spent a holiday. After two months her distraught parents arranged for a prayer meeting to be held in their home. Thereafter the poltergeist phenomena became less and less frequent, and died out altogether within a few weeks.

Childhood visions inspired the poet and painter William Blake.

Virginia Campbell and her home in Sauchie, Scotland – the scene of a famous poltergeist case in 1960.

Topham Picture Library

THE MIND

Strange Powers

It was a childhood poltergeist that first made the well known British psychic Matthew Manning aware of his supernatural powers. In Manning's case, however, these powers did not vanish with the disappearance of the poltergeist.

At 16, Manning was producing automatic writing, and paintings in the styles of Durer, Goya, Beardsley and Picasso. Inspired by Uri Geller, he then began bending metal objects and was studied by a group of scientists.

Geller's own talents also

Psychic Matthew Manning learned of his powers as a teenager.

showed themselves at a young age. At six, he could read his mother's mind, and often took a wicked delight in telling her when she came home from a card game exactly how much money she had lost! By 13, he was able to snap a bicycle chain through concentration alone and was using telepathy to pass exams. He says his technique was to stare at the back of a clever pupil's head and let the answers come to him!

Born Again?

Just as mysterious are the many cases in which children appear to retain a clear memory of a previous existence.

Dorothy Eady, born in London in 1898, was a perfectly normal

RITUALS OF CHILDBIRTH

Childbirth in the West is usually experienced in the security of a sterile maternity ward, but elsewhere it still carries considerable risks. Perhaps this accounts for the many strange rituals which have evolved among primitive societies to protect expectant mothers and newborn children from supernatural harm.

In many parts of the world the labour room is sealed to keep out malign spirits. In Greece it was once common to remove all mirrors to guard against the Evil Eye, while in some Italian peasant communities it is still the custom to sprinkle salt around the childbed to protect mother and baby against undesirable spirits.

On the Indian sub-continent, a knife hidden under the bed serves the same purpose. In Africa, the Azande tribe go so far as to pass newborn children over a fire to keep the forces of evil away.

Other primitive rituals concern disposal of the afterbirth. The Kwakiutl Indians of British Columbia threw a boy's placenta to the ravens, believing this gave him the power to see the future. A girl's placenta was buried on the beach to ensure that she grew up to be good at digging for clams – the Indians' chief source of protein.

The Azande tradition is to pass newborn children over a fire to ward off spirits.

toddler until the age of three, when a bad fall knocked her unconscious. During the blackout she dreamed of being in a strange and magnificent temple.

A year later, after a visit to the British Museum, Eady informed her family that she was the reincarnation of an ancient Egyptian priestess and that her true home was a temple in Abydos. This conviction never wavered, and when she grew up she moved to Egypt – eventually becoming the Keeper of the Temple of Isis.

There have been many cases of children as young as two describing events and people in a previous existence which have subsequently been proved to be accurate. Shanti Devi, born in Delhi in 1926, was seven when she told her family that her real name was Muttra and that she had been born before. She claimed she had been married with three children and died while giving birth to the last. She even took her parents to the house where Muttra had lived, where she recognized her 'husband' and two 'eldest children' but not the child who had cost her her life.

Superchildren?

The phenomenon of child prodigies may provide further evidence of reincarnation. Christian Friedrich Heiner, famous as the 'Infant of Lubeck', talked a few hours after birth, knew all the main events in the Bible at the age of one, was fluent in Latin and French at two – and at three had such an extensive knowledge of world history that he was invited to meet the King of Denmark. He accurately predicted his own death at the age of four.

Jean Cardiac also knew the alphabet at three months and could converse in English, Greek, Hebrew and Latin when he died at the age of seven. Other famous prodigies include Mozart, who was composing at four, and Pascal, who discovered a new geometrical system at 11.

Cases such as these are almost impossible to explain in rational terms. Is it possible that the children concerned brought the accumulated knowledge of their past lives with them, so that they could resume their work where they had left off?

BIRTH CYCLES

Recent scientific findings suggest that seasonal cycles and other cosmic rhythms have a significant effect on pregnancy and the first months of a baby's life.

For example, more births take place in the Northern Hemisphere in August and September, and babies born then tend to be stronger and more vigorous than those born at other times. American psychologist Florence Goodenough also found that children born during the summer had slightly higher IQs than those born in the winter. This has led to speculation of a link between fertilization and seasonal fluctuations in hormone secretion.

Research carried out in seaside towns to test the age-old belief that more babies are born when the tide flows than when it ebbs has proved inconclusive. But one interesting fact which did emerge was the sudden increase in births each day during the moon's passage overhead. This lends weight to the astrological argument that the moon's passage through the heavens has a major influence on birth.

THE MIND

THE EXORCIST

Can evil spirits really cause physical harm, and is there still a need to call on the traditional remedy – exorcism?

'*Demons ride like particles of dust in the sunbeam; they are scattered everywhere... they come down upon us like rain; their multitude fills the whole world, the whole air; yes, the whole air is a thick mass of devils...*'

The ideas of this 13th-century writer were common throughout the Middle Ages. Indeed, since the beginning of civilization, people have accepted the existence of a spirit world and believed that life is a never-ending battle against supernatural evil. Spirits of the dead and elemental beings were thought to be capable of taking possession of the home as poltergeists, while more menacing demons had the power to enter the human body.

To be bodily possessed by a spirit – whether good or evil – was thought to be one of the worst fates that could befall anyone. The signs were supposedly easy to recognize – victims lost weight, had fits and convulsions, screamed or shouted in strange voices, and possibly vomited items such as glass or pins. Medical opinion was rarely sought, and genuine illness never considered. Only one person was consulted in the search for a cure – the exorcist.

Battling with Demons

The word exorcism comes from the Greek *exorkizo*, meaning 'to conjure an oath'. This is the central part of the ritual used to banish an evil spirit from a person or place, following the example of

145

THE UNEXPLAINED

Jesus Christ, who, as the Bible describes, called on God to cast the Devil from a man who could not speak.

Armed only with salt, holy water, incense and a few sacred relics, both priests and laymen performed the rites using lengthy prayers whose form would vary depending on whether they were dealing with demonic possession or poltergeist activity.

When a person was possessed the exorcist would lead the victim into a church, sprinkling him with holy water and reciting prayers to banish the demon from the earth. Here, the true battle began. The spirits rarely seemed happy to relinquish their hold over their host and would cause the person to shrink from the exorcist, uttering blasphemies and snarls. Several helpers were often needed to hold the afflicted person down as he writhed and gnashed his teeth.

Exorcisms could take hours or even days to perform, for as one demon fled, another often took its place. The greatest danger to the exorcist was that of becoming possessed himself, since a priest was a prize indeed for a demon.

Spiritual Task

Exorcism reached its peak during the late 17th and early 18th centuries when the 'Witch Craze' swept Europe and America. The rituals became an almost everyday event and were performed on anyone whose behaviour did not

EXORCISM AND THE CHURCH

Exorcism is a subject of great controversy within the church. Although neither the Church of England nor the Catholic Church officially recognize the rite, many diocese have vicars or priests who will perform a 'blessing' if their bishop deems it necessary.

In 1963, the Church of England launched an official investigation into the matter and concluded: '... *every diocesan bishop should appoint a priest as diocesan exorcist...*' However, Dr Robert Runcie, then Bishop of St Albans, disagreed. During his reign as Archbishop of Canterbury, any questions on the subject were referred to him.

A screaming and writhing victim possessed by evil spirits. Demons were seldom happy to relinquish their hold over their host and would fight the exorcist at every turn.

THE MIND

At this 17th-century exorcism in Sainte-Baume, France, Catholic Inquisitor Michaelis is said to have cast 6,660 demons from the victim, Madeleine de la Palud.

satisfy those around them or comply with the accepted norm.

'Witches' who had summoned up devils and instructed them to enter a person or place were tortured and burned, while the methods used on the 'possessed' grew ever more elaborate. Many of these people, who were called 'demoniacs', were tortured, forced to suffer humiliating enemas of holy water, then beaten to drive out the Devil.

It was not until the mid-18th century and the dawn of the Age of Reason that the rituals began to fade into the superstitious past. As science flourished, victims of demonic possession became diagnosed as epileptics, schizophrenics or lunatics. Yet there were still some things – such as the occasional haunted house or incurable illness – that defied all logic. At such times, an exorcist would step from the shadows, using age-old rites to achieve results where more conventional methods had failed.

The Leeds Poltergeist

One of many such cases in modern times dates from 1982, when the Moore family were forced to call on the church to banish an unwanted spirit from their newly acquired home.

Six months after David, Janet and eight-year-old Jamie moved into the house on the outskirts of Leeds, the family decided to add a small conservatory to the side of the house. This was a simple enough job which David Moore planned to tackle himself. But on the morning after he began digging the foundations, strange things started to happen – literally with a bang.

At four in the morning a loud explosion woke Jamie, who ran to his parents' room, terrified by the noise. David and Janet had heard it too, but could find no obvious cause. Over the next few weeks bolted doors were found wide open, and Jamie's toys went missing, only to surface on high shelves in other parts of the house.

Then one night, a terrified Jamie came running into his parents' bedroom and told them about a strange man lurking in his room. Thinking it was a nightmare, David Moore escorted his son back to bed only to be struck by the room's icy chill. He, too, saw a shadowy figure, but convinced himself that it was only an hallucination.

ANIMAL EXORCISM

At the height of the Witch Craze, humans were not the only living creatures to undergo exorcism. Witches who 'cursed' people were thought capable of doing the same thing to animals. Beasts that suffered from an unknown illness, or perhaps just a little friskiness, were put through the same banishing rites as humans.

But why would the Devil wish to possess an animal? A reason was offered in Jerome's *Life of St Hilary*, written in about 390 AD: 'The Devil ... seizes even beasts of burden; he is inflamed by such intense hatred for men that he desires to destroy not only them but what belongs to them.'

On other occasions, exorcisms were performed to rid whole towns of pests. There are reports of vermin such as rats, mice, ants, bats, caterpillars and beetles, being successfully cast out after a ritual. Perhaps the tale of the Pied Piper of Hamelin was, in fact, an account of an exorcism.

The Pied Piper of Hamelin. Could the poetic tale in fact recount a long-forgotten animal exorcism?

THE UNEXPLAINED

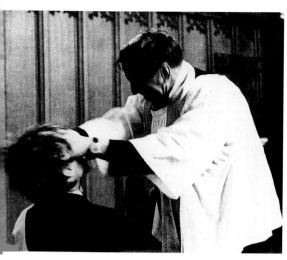

A modern exorcism. Reverend Christopher Neil-Smith of the Church of England dismisses a demon from a victim in 1971.

The last straw came when Janet Moore returned from her part-time job the following day to find bookshelves overturned and their contents scattered about the house. A police investigation revealed that nothing had been stolen and there was no sign of a forced entry.

A neighbour finally moved the family to action after confessing that she was receptive to psychic phenomena. She said she had already sensed something was wrong, and when shown where David Moore had been digging, insisted there was a strong feeling of evil about the place.

The neighbour put the family in contact with a local priest who visited the following day. After ordering all the windows to be opened, he moved from room to room reciting prayers and sprinkling holy water. Finally he blessed the Moores themselves. The family were never troubled again.

David and Janet Moore later discovered that their housing estate had been built over an old crossroads – a popular spot for hangings in the 17th century. They now believe the digging may have awakened a malevolent spirit. Even today, David is puzzled by the incident: 'Something was there, and now it's gone. The exorcism worked, but how I've simply no idea. I'm just grateful that it did.'

THE DEVILS OF LOUDUN

On the 18th April 1634, a priest named Urbain Grandier was burned at the stake after being accused of bewitching a whole convent of Ursuline nuns in Loudun, France. In the months before, the nuns had shrieked, writhed, screamed obscenities and performed lewd acts. Grandier, a prosperous and unorthodox priest, was called in to perform an exorcism but was later accused of invoking the evil spirits himself. Some believed that jealous enemies had asked the nuns to fake hysteria.

After Grandier's death the case took a bizarre twist. Although the demonic possessions should have ceased, the nuns continued their frenzied antics, and other priests linked with the incident experienced similar fits. Three of them died within four years – one claiming to have seen Grandier's ghost. The fourth spent the next 25 years suffering bouts of what he called 'demonic possession'.

Was it simply mass hysteria or was the Devil's work responsible? The case will always remain open.

*The controversial film **The Devils** retold the story of the nuns of Loudun. Was there a conspiracy against Grandier?*

THE MIND

TORMENT AND TRANCE

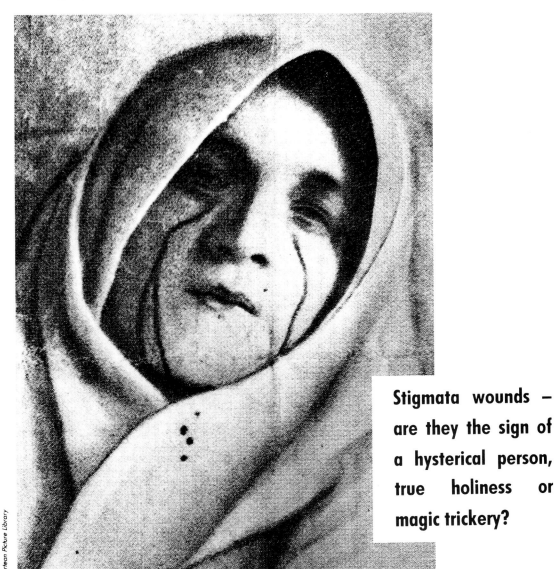

Stigmata wounds – are they the sign of a hysterical person, true holiness or magic trickery?

At first, 12-year-old Cloretta thought it was sweat that was pouring from her brow. It was certainly very warm in the classroom that day, and the mathematics lesson required a lot of concentration. Taking a handkerchief from her pocket, she mopped her forehead and was never so shocked in her life. She was bleeding profusely, even though she was unaware of having scratched herself. The whole class – including her teacher – were witnesses to the fact.

Stigmata

Baptist by faith, this highly religious girl from Oakland, California, first experienced stigmata (the spontaneous appearance of wounds similar to those inflicted upon Christ on the cross) in 1974, and they reappear each year, round about Easter time. She is now a healer.

That episode in the classroom was, in fact, the culmination of a series of events that had occurred over the previous few days, when Cloretta had suddenly begun to bleed, without warning

THE UNEXPLAINED

Padre Pio of Italy is thought to have suffered stigmata on his hands, feet and chest for over 50 years.

of any kind, from either her hands or feet.

Several doctors were invited by Cloretta's naturally very anxious family to examine her, but all found her in good health, and were unable to give a satisfactory explanation for the extensive bleeding. What is more, the blood was definitely of her type and grouping, so the possibility of deception was discounted. Yet why should this dreadfully stressful experience happen to her? Why, too, over the last few hundred years, whenever similar phenomena have first occurred, has it almost always been at Easter and usually on Good Friday itself?

Cycle of Suffering

Some victims have experienced stigmata with terrifying frequency. Gemma Galgani of Italy, for instance, who subsequently became acknowledged as a saint, suffered for years in this way, after being cured of tuberculosis at the age of 25. The stigmata would occur like clockwork at precisely 8pm every Thursday.

Therese Neumann of Bavaria, who had been born on Good Friday, suddenly noticed stigmata on her hands and feet on her 28th birthday, and thereafter on every birthday. (She also claimed to have lived for years without food of any kind.)

Louise Lateau, a Belgian peasant girl who lived in the 19th century, had similar experiences while in a trance-like state, on many hundreds of occasions before she died. The Belgian Academy of Medicine, however, could find no rationale for her regular haemorrhaging, even though they experimented by encasing the arm that would bleed within a glass container, thereby preventing outside interference and presumably eliminating trickery of any kind. Examination of the sites of her wounds under a microscope revealed only the tiniest triangular marks, but the bleeding was always heavy.

Saintly Scars

Wounds of this sort were first described by St Francis of Assisi in the year 1224. Since then the devout of the Catholic church in particular have regarded their appearance as an outward sign of saintliness. To date, approximately 300 cases have been re-

FAKIRS

For centuries the fakirs of the Indian sub-continent have been performing amazing self-inflicted tortures. These include passing metal skewers right through facial flesh, seemingly without agony; swallowing broken glass unhesitatingly; and lying prone on a bed of nails while almost naked.

Some fakirs are of the Islamic faith, others Hindu. But the sort of self-mutilation and bodily stress that they practise is uniformly presented as part and parcel of an attempt to achieve a holy state through the renunciation of worldly pleasures and the seeking of spiritual powers.

Some will stay rooted to the spot for years on end; others opt to stand in ice-cold, waist-high water for long periods of meditation; one sect even puts a ring through the penis as a sign that they have overcome both pain and sexual desire.

Certain of the fakirs' 'tricks', it is claimed, can actually be learnt. Others believe, however, that fakirs really do endure intense pain and physical stress through sheer mind power.

Experiments show a change in the brain wave patterns of fakirs practising intense concentration, which may reduce pain.

corded. Interestingly, too, a number of Moslems have experienced stigmata that are said to resemble the wounds that Mohammed met with on the battlefield. Some, however, have sought to prove them to be fraudulent.

Sometimes, very heavy bleeding is reported; sometimes there are raw wounds or warts that suddenly appear; others do not haemorrhage at all but feel acute pain in those areas of the body where Christ is thought to have bled from his wounds.

Generally the scars are said to disappear completely; but throughout the later years of his life, Padre Pio – the famous priest of San Giovanni Rotondo in Italy who would often go into a trance while conducting Mass – had to wear exceptionally wide shoes in order to accommodate the bandages he wore to protect his sore and bleeding feet. The Vatican eventually recognized his sufferings as a miracle and forbade him to leave his monastery, but did permit him to build an adjacent hospital with donations made by visiting pilgrims.

Wounds of Hysteria?

A great many explanations have been offered by those who are sceptical about the cause of stigmata. Some have suggested that the wounds are definitely fakes; others that they are self-inflicted during an hysterical trance or even as a deliberate attempt to become recognized as a saint. A certain nun, for instance, suffered from a form of nervous behaviour which involved constant rubbing of her fingers and toes. She bled regularly, yet others saw the resulting wounds as evidence of a miracle – and so did she.

Sheer hard physical labour, according to some, may well

St Francis of Assisi was said to have received stigmata after a seraph appeared to him in a vision.

PSYCHIC SURGERY

Some people seem to have the gift of psychic healing (see page 77). They can improve a patient's health or even cure him or her completely by the laying on of hands, through prayer and meditation, via samples of hair and nail clippings (as in radiesthesia), or through various forms of telepathy or thought transference.

The psychic surgeons of the Philippines and South America, however, go one step further and are known to carry out actual operations – for the removal of tumours, for instance – without the use of instruments or pain relief of any kind. It might be expected that total lack of anaesthesia would cause the patient intense physical and mental distress, yet this does not seem to be the case.

During a psychic operation, the 'surgeon' appears to open up the patient's body simply with his hands. Almost magically, he will then produce lumps of fatty looking tissue which are said to be the cause of the complaint, whatever its nature. And when he has run his hands over the area where he has 'operated', no trace of a scar will be visible.

Those suspicious about the skills of psychic surgeons remain convinced that because the patients believe they had an operation, they feel better. And some think that the tissue produced as evidence of the operation having taken place is merely the result of a clever conjuring trick. Careful scrutiny of films that have been made provides little by way of conclusive evidence either way: but what is certain is that many of the patients who undergo psychic surgery do affirm a cure.

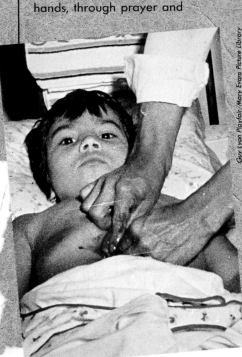

Psychic surgery is a popular form of healing in Brazil and the Philippines where apparently successful operations are performed on various parts of the body, such as the eyes and stomach. In the West, people remain sceptical about it.

Stigmata are said to appear on those parts of the body where the crucified Christ was wounded.

have caused what were apparently signs of divine favour in certain members of holy orders.

Painful Memories

Psychiatrists have also described cases in which just the very memory of some dreadful past experience has resulted in the appearance of ulcers or dermatitis: the scars of the mind, it is suggested, can show up on the body, too.

It has also been shown that wounds and bruises can be made to appear at will by some people. A Swedish girl, for example, who was once very severely beaten up, would bleed whenever she got very excited or angry – a reaction that is actually more often reported as taking the form of severe bruising. Some researchers have also considered the possibility that it may be pink-coloured sweat that is produced, caused by blood seeping into the sweat glands.

Hypnotic Suggestions

Since the late 19th century, a number of experiments have shown it is possible to bring about certain skin conditions – blistering and perhaps also wounds that resemble those of Christ – by means of hypnotic suggestion. It has been pointed out, too, that many famous depictions of the crucifixion are in fact inaccurate, since the Roman method involved driving nails through the wrists only. In other words, victims most often experience stigmata where they think Christ had his wounds and not where they are far more likely to have been inflicted.

THE MIND

THE
Miracle Man
CASEBOOK

Indian mystic Sai Baba materializes objects out of thin air with a wave of his hand. How?

In a materialistic and cynical age, is it any wonder that people the world over are looking for a new direction? Some involve themselves in environmental issues or charity work; others see religion as a potential solution. In their search for enlightenment, millions of people from all parts of the world are devoting themselves to the teachings of a mysterious Indian religious figure and mystic – Sathya Sai Baba.

Distinguished only by his bright cotton robes and bushy hair, Sai Baba preaches hope, peace and love. He claims to be one of the latest in a long line of Hindu *avatars* (incarnations of gods), born to take man from the darkness of the machine age into the light of spiritual knowledge.

This holy man's philosophy touches on politics, social welfare and education. He has established schools, colleges, a university, a hospital and organized groups of volunteer workers. However, it is his ability to perform miraculous feats which continues to astound his devoted followers. Every year, thousands of people flock to Sai

153

THE UNEXPLAINED

Baba's famous ashram Prashanthi Nilayam (Abode of Great Peace) in the southern Indian town of his birth — Puttaparti, in the state of Andhra Pradesh. Few leave without observing the miracle man in action.

Amazing Powers

With a simple wave of his hand, Sai Baba produces *vibhuti* (sacred ash), sweets, coins, religious medallions and pieces of jewellery which he distributes among his adoring public. Some claim to have eaten food – both hot and cold – pulled out of thin air by the holy man or conjured up from limited supplies. Baba jokingly refers to the mysterious dimension from where he draws these apports as the 'Sai Stores'.

Some older devotees believe they have seen Sai Baba teleport and levitate. Others say that he can read their minds. He is also credited with several miracle cures and is even said to have raised two people from the dead.

Elsie Cowan claimed her husband Walter, a wealthy Californian businessman, died in her arms on Christmas Day in 1971. Allegedly, she went to beg Sai Baba for help and he travelled to the hospital where Walter lay. What happened there is unrecorded, but when Elsie arrived later her husband was alive again. In another case, Sai Baba apparently revived factory owner V. Radhakrishna in 1953. His body had shown no signs of life for 20 hours.

Early Promise

Curious tales like these have surrounded Sai Baba since his birth in November 1926. His parents named him *Sathyanarayana* – *Sathya* meaning truth and *Narayana*, deity. When he was a baby, his mother and father found him one day rocking in his cradle, nestled safely within the coiled body of a cobra – a Hindu sign of divinity. Stringed instruments which hung in the young child's room are also said to have played music of their own accord.

The holy man — with trinkets allegedly pulled from thin air.

One of Sai Baba's many colleges in India, the land of his birth.

A popular child at school, Sathyanarayana led a worship group where he delighted friends by producing sweets and gifts from what appeared to be an empty bag. The only thing that disconcerted his parents was their kind-hearted son's habit of bringing home beggars to be fed. After they challenged him about it, the boy often chose not to eat himself – giving food from his own plate to the poor.

One day when 13-year-old Sathyanarayana was walking with friends, he suddenly leapt into the air with a shriek. His friends feared he had been bitten by a scorpion, but Sathya showed no signs of illness until he collapsed later that evening.

Doctors could not revive the boy, who astounded everyone by awakening as if from a nap the next day. From then on he was a different child. He would suddenly burst into song, or recite poetry and long religious passages in the ancient language of Sanskrit – texts which he had never (and could never have) learned.

His alarmed parents called in

THE MIND

THE LONG LINE OF AVATARS

According to the Hindu scriptures, an avatar is an incarnation of a god who chooses to be born into the world of men to help them.

The best known avatar is Vishnu, who, according to Hindu mythology, has already lived among men nine times. He has appeared as a fish (Matsaya), a tortoise (Kurma), a boar (Varaha), a half man-half lion (Nrisinha), a dwarf (Vamana), an axe-bearing hero named Rama (Parasurama), Rama again (Ramachandra), Krishna and finally as Buddha. Legend has it that he will become manifest one more time as a white winged horse (Kalki) that will finally destroy the world.

Avatars are said to exhibit exceptional qualities which set them apart from men. All-powerful and all-knowing, they are also able to materialize and disintegrate matter at will and radiate divine love. Since they are not bound by the laws of *karma* (cause and effect), they have the unique ability to transform the lives of mankind.

Sai Baba is an avatar because he claims special powers inherited from his past life as a holy man. He predicts that he will live to the age of 94, dying in 2020 or 2021 – to be reborn again as Prema Sai in order to continue his mission.

Vishnu in an early incarnation as the tortoise (Kurma).

doctors and, as a last resort, an exorcist. But it was to no avail. Instead, the boy called the family together and with a wave of his hand, produced sweets and flowers for them. His angry father asked him if he was a ghost, a god or a madman. 'I am Sai Baba,' said the child, 'come to ward off your troubles and keep your houses clean and pure.'

His parents had never heard of Sai Baba. However, older people in the village remembered the former Hindu holy man – a performer of miracles – who often healed the sick with sacred ash. Before his death in 1918, he had promised to be reborn in order to continue his mission.

The original Sai Baba – Sathya in a previous incarnation?

Trick or Treat?

With so many apparently inexplicable stories surrounding Sai Baba, it is hardly surprising that his life has been dogged by controversy. Critics have accused him of using sleight of hand, saying that he pulls apports from his sleeves or hides them in his hair. His refusal to allow scientists to test his incredible talents in a laboratory has left him open to further accusations of fraud.

Yet researchers and journalists who have observed his feats closely have never caught him engaging in trickery of any kind. Psychic researcher Dr Erlendur

THE UNEXPLAINED

Haraldsson met Sai Baba for the first time in 1973. With a wave of his hand, the mystic produced a rare double *rudraksha* – an acorn-like object which resembles a piece of twinned fruit. Haraldsson admired it then passed it back to Sai Baba who enclosed it in his hands and blew on them. He then revealed the rudraksha which was now in a gold setting with a chain. He presented the token to the researcher as a gift.

During his visits, Haraldsson noted that Sai Baba produced as many as 50 trinkets a day. His long, cotton robes appeared to contain no pockets, nor could he hide such a vast number of objects in his hair. Stranger still, many devotees received tokens which they had requested specifically. 'Magicians cannot produce objects by sleight of hand without prior preparation,' Haraldsson observed.

But the researcher also expressed reservations. As he pointed out, trickery cannot be ruled out

Some claim they have seen Sai Baba take on the form of Krishna.

COLOURS OF LOVE

Many people claim they can see the human aura as an emanation of colours which radiate from the body. The colours displayed are said to reveal many things about a person's health, personality, emotions and state of mind.

University of Arizona professor Dr Frank Baronowski – a gifted clairvoyant – believes Sai Baba's aura surpasses that of any holy man he has studied. He saw gold, blue, pink and silver radiating from Baba's body. Students of auras believe that blues indicate intuition, pinks show love, while gold and silver reveal spiritual purity.

Sai Baba's aura of love captured by psychic artist June Lockhart.

British psychic artist June Lockhart from Surrey has also met Sai Baba. She remembers: 'As he stepped through the door of the mandir (temple) I witnessed an aura quite unlike anything I had seen before. The intensity and magnitude of Baba's energy field seemed to embrace the vast gathering of people with a dynamic love beyond description. I could never find a canvas or paint adequate enough to capture what I saw – though I have tried.'

merely because he did not see it. He was also disappointed that he could not convince Sai Baba to perform under laboratory conditions but understood the Indian's reluctance to submit to another culture's tests.

To this day, Sai Baba continues to give the same answer to his critics. He says that his ability to perform miracles is a sign of divinity – a simple calling card to attract followers who will later hear and learn his spiritual message. 'I give you what you want in order that you may want what I have come to give,' he explains. To many, Sai Baba's feats may look like 'magic', but much more important is his promise of hope and peace — a ray of sunshine amidst the clouds of darkness.

UNSOLVED MYSTERIES
LOST AT SEA

Can anything account for mysterious sea disappearances such as those which have occurred within the Bermuda Triangle?

'We are not sure where we are. We think we must be 225 miles north-east of base. It looks like we are entering white water . . . Don't come after me or . . .' Crackles of static and finally silence marked the end of the last radio message from Flight 19. It was 4pm on December 5th 1945, only two hours after five US Navy TBM-3 Avenger torpedo bombers had taken off from Fort Lauderdale, Florida on a routine training exercise

A search operation was quickly mounted in the few hours that remained before dusk, focusing on an area north of Grand Bahama Island. Within half an hour, a Martin Mariner flying boat with 13 people on board had also vanished without trace.

The massive searches which were organized over the following days covered an area of 980,000 square kilometres, yet no clue as to the fate of the 27 crew of Flight 19 or the Martin Mariner has ever emerged. These losses, together with that of the crew of the *Mary Celeste*, are the most celebrated of some 200 recorded disappear-

THE UNEXPLAINED

ances from the now legendary Bermuda Triangle.

Eerie lights, malfunctioning navigational equipment and sudden disappearances have long been linked with the Bermuda Triangle. For years it seemed that such strange phenomena were confined to a section of the Atlantic Ocean bounded by Florida, Bermuda and Puerto Rico, but the latest reports suggest the Triangle covers a much wider, less clearly defined area.

One of the most bizarre tales concerns a large glowing white disc which rose from the ocean just north of Cuba in October 1969. The event was witnessed by the entire crew of an American guided missile destroyer, the *DLG-27*, who were unnerved when the object failed to register on their radar screens.

Mystery Malfunctions

There have also been many reports of ship's compasses going haywire in the area. In February 1955, for example, a submarine icebreaker, the *USS Tigrone*, sailed more than a mile off-course and collided with a reef despite having the benefit of advanced navigational equipment. In March 1971, another US Navy ship, the *Richard E. Byrd* was stranded between Norfolk, Virginia and Bermuda for ten days with all communications cut. And since then, several ships have reported a sudden, inexplicable loss of power after passing through some mysterious kind of sea mist.

The Richard E. Byrd – stranded at sea for ten days with all communications cut.

Five TBM-3 Avengers disappeared over the Bermuda Triangle near Grand Bahama Island. What was the fate of the crew of Flight 19?

The largest ship to vanish in the Triangle was a 17,560 ton collier, the *Cyclops*, in 1918. Yet disappearances are not confined to ships, planes and their crews. In 1969 two men were reported missing from the Great Isaacs Lighthouse in a loss almost identical to one from the nearby Flannan Islands Lighthouse at the turn of the century. Several scuba divers have also vanished, including Archie Forfar and Anne Gunderson, who were lost without trace by their five-man back-up team during a world diving record attempt in 1971.

UNSOLVED MYSTERIES

Of the many theories put forward to explain the mysteries of the Bermuda Triangle, perhaps the most intriguing concerns interference by extra-terrestrials. The author Charles Berlitz claims that the last phrase heard from Flight 19 was '... *it looks like they are from outer space*'.

Other experts have speculated on the presence of some kind of terrestrial 'black hole', where changes in the earth's gravitational field suck in anyone or anything nearby. A few also speak of possible contact with a 'Fourth Dimension' in which the conventional barriers of space, time and distance break down.

The naval supply ship Cyclops – the Triangle's largest victim.

Fact or Fantasy?

Evidence in support of the 'time warp' theory comes from an incident dating back to December 4th, 1970, when Bruce Gernon Jr and his father watched their light aircraft enter what they later described as a 'vertical tunnel' within a bank of cloud. The plane suffered a temporary loss of gravity and flew into a green fog, whereupon the flight instruments ceased to function. After flying through one of several widening gaps in the fog, they reached their destination 30 minutes ahead of schedule with only 28 of their expected 40 gallons of fuel used.

Among the more earthbound theories put forward to account for individual disappearances are piracy, fatal disease, attack by giant squid, and human error coupled with a lack of safety precautions. It seems, however, that many incidents can be explained by the notoriously freak weather conditions experienced around this part of the ocean.

Tropical storms, hurricanes, tornadoes and tidal waves are all

THE MYSTERY OF THE MARY CELESTE

On December 4th 1892, 400 miles west of the Azores, the captain of a Gibraltar-bound ship, the *Dei Gratia*, hailed a vessel which appeared to be sailing somewhat unsteadily. What the captain and his mate discovered aboard the brigantine *Mary Celeste* later that afternoon has fuelled intense speculation ever since.

The ship was eerily deserted, and there was no obvious sign of a struggle. All money and valuables were still locked away in the safe, clothing and personal possessions seemed to be untouched, and even the dishes had been put away after the last meal. Later findings established that the *Mary Celeste*, which had left New York bound for Genoa, was in fact sailing westwards, some 400 miles from its original course. Whatever could have happened to the captain and his nine crew?

Since the mystery first caught the public's imagination, the men's disappearance has been blamed on anything from mutiny to the activities of an irate sea monster. Researchers have shown, however, that the true facts of the case quickly became obscured – first by an incompetent official investigation, and then by the novelist Arthur Conan Doyle's fictional explanation; his fantastic account of the story published under the title the *Marie Celeste*, was later accepted as fact by many writers.

What seems certain is that the crew left the ship of their own accord, since a longboat was found to be missing along with the captain's instruments and charts. Was their sudden departure the result of hitting some kind of oceanic whirlpool as one recent investigator has tried to prove? It seems unlikely that the ocean will ever give up one of its most famous and best-kept secrets.

Captain Benjamin Spooner Briggs vanished with the Mary Celeste.

THE UNEXPLAINED

relatively common, while fireballs – ball lightning – can travel through glass and metal and are known to cause severe radio interference. There is also evidence to show that areas of change in the earth's electromagnetic field – marked as Doldrums on sea charts – can affect the human nervous system, resulting in illnesses such as hallucinations or migraine.

As scientific research into the effects of electromagnetism and freak weather conditions is stepped up, it may only be a matter of time before the bulk of Bermuda Triangle disappearances receive rational explanations. In the meantime, the fate of Flight 19 and the other losses from the area remain as baffling as they have ever been.

Dutch investigator of the paranormal Gerard Croiset points to a map of the Triangle as he discusses theories to account for the mysterious disappearances.

MONSTERS OF THE DEEP

Sea charts until the 18th century were invariably decorated with fanciful monsters ranging from winged sea-dragons spouting water to sea serpents with undulate bodies. Belief in such creatures was fostered by a brisk trade in skate and rays faked to resemble 'dragons' and in 'mermen' made from dried monkeys and fish tails.

However, perhaps the stories contain a grain of truth. In 1848 the captain and crew of the *Daedalus* testified to having seen a 18 metres (60 feet) sea serpent swimming at a rate of 15 knots off the Cape of Good Hope. The creature was described as having a snake-like head which on occasions reared four feet out of the water.

Recent reports continue to baffle the public. The most famous is the *Morgawr*, first sighted in September 1975 off Falmouth in Cornwall and seen again many times during the summer of 1985. The creature was described by one witness as dark in colour, with a snake's head and a humped body revealing skin like a sealion. An unknown species of whale, perhaps – or a monster from the deep?

◀ *The Morgawr – photographed off the coast of Cornwall in 1976.*
▼ *A sea serpent claims its victim.*

UNSOLVED MYSTERIES

DEADLY CONTACT
CASEBOOK

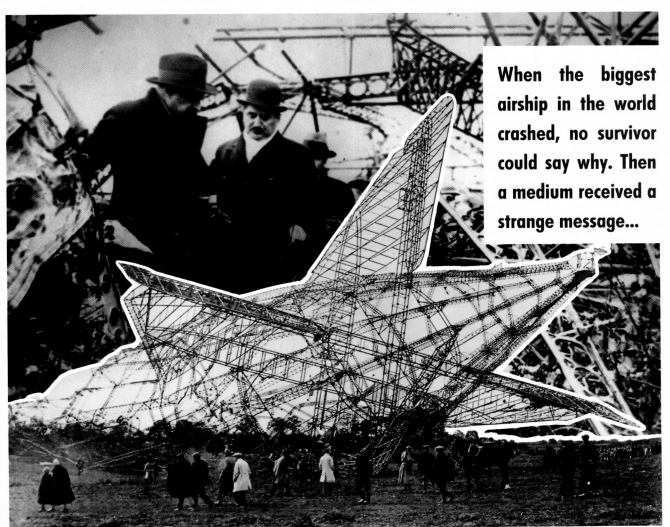

When the biggest airship in the world crashed, no survivor could say why. Then a medium received a strange message...

On October 5th 1930, the British public were stunned by news of a horrific air tragedy. The world's largest airship – the R101 – had crashed into a field near Beauvais in northern France while en route to India. It exploded at once into a ball of flames. All but six of the 52 passengers and crew died in the blaze.

It was the age of airships, but controversy had surrounded this ill-starred vessel from the outset. Even while the R101 was being built at the Bedford suburb of Cardington, rumours circulated about possible design flaws. The diesel engines seemed too heavy; the outer fabric, flimsy; and the internal gasbags and an experimental fuel, unreliable.

Dark Rumours

Worse was the talk that Lord Christopher Birdwood Thomson, the then Secretary of State for Air, had possibly used the craft to further his own political ambitions. He had set an impossible deadline for the airship's completion, so it could make a highly prestigious flight to India. Had the safety trials been rushed?

While the tragedy at Beauvais appeared to confirm many dark suspicions, the truth lay buried with the flight officers, who had all died in the crash. The story might have ended there, but for a dramatic development. Eileen Garrett, a highly reputed Irish-born medium, unintentionally made contact with the spirit of the dead flight captain!

161

THE UNEXPLAINED

The first spirit message was received through Eileen at a seance held in London less than 48 hours after the R101 crash. Harry Price, the British ghost-hunter, was attempting to contact his friend Sir Arthur Conan Doyle – the author of the Sherlock Holmes mysteries – who had recently died. The two men had been sparring partners: Conan Doyle was an advocate of spiritualism while Price was its fiercest debunker. Journalist Ian Coster attended as a witness.

Instead of Conan Doyle, however, Harry Price heard the voice of 'Bird' Irwin, the late Captain of the R101, saying urgently: 'Engines too heavy... useful lift too small... elevator jammed... oil pipe plugged.' The complaints streamed out. Tears stained the medium's cheeks and her hands clenched and unclenched as she continued: 'Cruising speed bad and ship badly swinging... cannot rise... same in trials... no one knew the ship properly.' Price and Coster sat spellbound. Neither of them knew much about airships, but the welter of technical jargon convinced them that the medium could not have invented the messages. Nor could she have gleaned the information from news reports. The Air Ministry was maintaining a guarded silence on the subject until an official inquiry was completed.

Ghostly Message

At the Court of Inquiry, which opened on October 28th, there was soon cause for concern. Crucial documents relating to the R101's flight tests had been lost, and there were widespread rumours that the Ministry was attempting a cover up.

One man at the hearing had a good reason to feel dissatisfied. A

▲ *The once mighty R101 was quickly reduced to a mere skeleton after its explosion in France.*

▶ *A message received by Irish medium Eileen Garrett (inset) provided the first clue as to what had really happened.*

UNSOLVED MYSTERIES

Court of Inquiry: public whitewash?

> ### THE PREDICTION
> Even before the R101 embarked on its fateful trip, there were portents of disaster. The strangest was a posthumous warning from Captain Walter Hinchcliffe who had disappeared in 1928, while trying to cross the Atlantic Ocean in an airship called the *Endeavour*. His widow, Emilie, began visiting a medium at the London Spiritualist Alliance, and started to receive messages from her dead husband.
>
> Most communication related to the couple's domestic affairs, but in August 1930, the topic changed to the R101. Hinchcliffe asked his wife to warn his friend, the R101's navigator Squadron Leader Johnston. 'I am afraid they are rushing things. It will not be a success,' he apparently said. In September, the messages became more urgent: 'The vessel will not stand the strain... it will come down.' Emilie Hinchcliffe made two trips to the R101's base at Cardington to try to convince Johnston – but it was to no avail. On the day of the flight she heard her husband's sombre comment: 'Storms rising... nothing but a miracle can save them.'

Mourners file past the victims at Westminster Hall. Captain Irwin and his ill-fated crew (above).

few days before it began, Major Oliver Villiers, an Intelligence Officer in the Air Ministry, had had a peculiar experience. While he sat drinking a cup of tea alone, he sensed that someone was sitting beside him. In his mind, he heard his old friend Captain Irwin say: 'For God's sake let me talk to you. We're all bloody murderers.'

Alarmed, Villiers decided to consult a spiritualist. By coincidence a friend put him in contact with Eileen Garrett, the same medium that Harry Price had used. At a seance held on October 31st, Villiers kept his identity and profession a secret. When Captain Irwin's voice came through, Villiers immediately recognized its Irish lilt.

THE UNEXPLAINED

Irwin's tone was compelling. After repeating the description of the crash, he warned his friend against testing the R101's sister ship. 'Over here, we call her the inventor's nightmare. She is all wrong in construction,' he said. Irwin also complained about the Court of Inquiry whitewash: 'It is dreadful to hear the things they are saying – all bosh and they know it.'

The R101's captain was not the only spirit voice to comment on the disaster. Villiers met seven times more with Eileen Garrett and 'talked' to several crash victims, including the R101's navigator, Squadron Leader Johnston. Another medium from Bedford, Bertha Hirst, also received messages from the Squadron Leader at a seance attended by his wife. The widow was convinced that Hirst's description of the spirit of a man holding a white rose was her husband – the flower had been his parting gift to her before he embarked on the ill-fated flight.

Spirit Truth

No rational explanation was ever found for the supernatural messages. Cardington officials examined a transcript of Captain Irwin's 'conversation' with Harry Price, and accepted it as a more plausible account of the crash than the vague findings which had emerged from the public inquiry. These experts claimed its technical accuracy ruled out the chance of a spiritualist fraud.

Irwin's spirit, for example, spoke of the experimental fuel – a combination of hydrogen and oil — referring to a top-secret experiment known only to a few technicians who had been closely involved in the project. He also described how the R101 had 'almost scraped the roofs at Achy'. Investigating, Price found that while ordinary maps did not show the tiny hamlet near Beauvais, it featured on the pilots' large ordnance charts.

Villiers had to wait almost 40 years for his own beliefs to be confirmed. In 1967 the widow of the R101's First Officer, Lieutenant Commander Atherstone, finally released her husband's diary. In a seance held shortly after the disaster, Villiers learned that it contained critical information relating to the crash. At last, he could read the airman's comments: 'There was a mad rush and panic to complete the ship... it has no lift worth talking about and is tail heavy.' To this day, the R101 mystery remains just as tantalising as ever.

THE SAGA CONTINUES

The disastrous crash at Beauvais was not to be the final chapter in the history of the luckless R101. The twisted metal girders from the airship were shipped back to England where they were auctioned as scrap. The successful bidder was the Zeppelin Company in Frankfurt, which incorporated much of the waste metal into the framework of its new airship, the *Hindenburg*.

The craft seemed to fulfil the promise of the R101. Over 80m (264ft) long, it had the capacity to transport 72 passengers in comfort across the Atlantic in just 52 hours. One satisfied traveller said the trip was: 'As motionless and luxurious as if we had been carried through space on a magic carpet.'

Despite the worsening political climate brought about by the rise of the Nazi régime in Germany, commercial flights were operating by 1936. But the new craft had only made ten round trips when disaster suddenly struck.

On May 6th 1937, the mighty *Hindenburg* was about to land at Lakehurst, New Jersey when there was a huge explosion. A hydrogen leak had ignited and the tail of the craft immediately burst into a giant column of yellow flame. Moments later, it plummeted to the ground killing no fewer than 35 people. Had the jinx of the R101 claimed a new set of victims?

Great balls of fire consumed the Hindenburg, an airship built with metal taken from the R101.

UNSOLVED MYSTERIES

CURSED FOREVER?

Do some treasures from the past have the power to put a killing curse on their hapless owners?

Cursed possessions which bring misfortune or death upon their owners have long been the lifeblood of horror stories. Can inanimate objects really cast an evil eye over those who covet them? Or does their apparent power reside firmly in the minds of a few guilt-stricken individuals?

While cursing today implies little more than bad language, there was a time when curses were almost universally believed to invoke spells bringing bad luck or even death. Many were simply crude instruments of vengeance, directed against those who had taken what they wanted by force. But curses played an important protective function too; by reinforcing existing social taboos, they allowed both Church and State to keep their most treasured possessions in safe hands.

Perhaps the best example of a 'protective' curse surrounds the fabulous wealth which is said to lie buried with the Pharaohs of Ancient Egypt. Ever since the first tombs were opened and rifled by Western explorers, stories of cursed objects which pursue their victims beyond the grave have become legion.

One of the most celebrated curses from the land of the Pyramids concerns the lid of an apparently 'haunted' sarcophagus

THE UNEXPLAINED

owned by the British Museum. Discovered in the 1860s, this mysterious relic was believed to be the burial casket of a Theban priestess of Amen-Ra and reputedly brought bad luck to all who came in contact with it.

The story goes that one of the lid's first owners, Douglas Murray, lost an arm in a shooting accident soon after he purchased it. The lid was then borrowed by a London lady journalist and in rapid succession, her mother died, her engagement was broken, and she contracted a mystery illness.

The Mummy's Curse

Tales of the macabre jinx persisted after Murray donated the lid to the British Museum. An Egyptologist reportedly died while studying the inscriptions, and a photographer produced film which he claimed showed the peaceful face on the outside of the lid contorted with malice.

By the 1930s the sarcophagus had acquired an almost legendary reputation, and donations flowed in from around the world along with requests that floral tributes be placed near the display. The Museum felt obliged to deny all rumours of a curse, but other stories of sudden deaths at the sites of archaeological digs continued to fuel speculation that the contents of the Pharaohs' tombs lay under the protection of a sinister curse.

Engaging though these stories are, it seems likely that they are entirely fanciful. Egyptologists now believe that the fearsome curses inscribed on the burial chamber walls in the Pyramids constituted no more than a warning to those responsible for the upkeep of the tombs. Science, meanwhile, has linked the mysterious deaths following entry into the tombs with the presence of unknown bacteria that may have lain dormant for centuries.

Haunted Skulls

Other tales of bewitched objects seem to defy rational explanation. Among the many stories of gruesome relics which have brought misfortune on their owners are several cases of 'haunted' skulls recorded by ghost-hunter Robert Thurston Hopkins.

One famous tale involves the Victorian 'Red Barn' murderer William Corder, who did away with his lover in Polsted, England, in May 1827. Some 50 years after his execution the preserved remains of Corder's body were acquired by a certain Dr Kilner. The doctor duly severed the skull from its skeleton, polished the pickled scalp to a gloss and proudly exhibited the macabre trophy in his surgery.

Kilner also prided himself on his detached, scientific view of the paranormal, yet nothing could prevent him from being severely disturbed by the events which followed. First, both he

The British Museum's celebrated sarcophagus lid – the subject of an Ancient Egyptian curse?

and his maid saw a stranger wearing Victorian clothing near the surgery, and he became plagued by breathing and muttering sounds. Finally, he awoke one night to crashing noises in the room below.

Descending the stairs, Kilner was met by an icy gust of wind. He went on to find the display cabinet open, the box which had contained the relic smashed, and the skull itself magically transported to a nearby shelf from which it appeared to be grinning malevolently at him.

After a later search revealed no signs of human entry, the terrified doctor passed on the remains to Thurston Hopkins' father, who gave them a Christian burial. The disturbances ceased.

The Strangler Jacket

Skulls are not the only objects to attract mysterious dark forces. Even something as simple as a piece of clothing may take on a frightening jinx, as in the case of the Victorian bolero-style jacket owned by The Duke of York's Theatre, London, which was later dubbed the 'Strangler Jacket'.

In 1948 the celebrated comic actress Thora Hird appeared at the theatre and wore the jacket as part of a costume for a period play. After a time she was terrified to discover that it grew tighter with each performance to the point where she felt it was smothering her.

When Hird's understudy had the same experience, the two women asked other members of the cast to try on the jacket – with similar results. One man almost passed out, while the director's wife was left with angry, red marks around her throat.

Precious little was known of the garment's history and there appeared to be no explanation for the group's experiences. It was not until a seance was held at the theatre that the jacket was linked with the brutal murder of a young girl who had been drowned in a

Red barn murderer William Corder – his skull haunted its owner.

THE CURSE OF TUTANKHAMUN

Excavations at Tutankhamun's tomb provided a celebrated instance of an ancient curse brought into the modern world. George Herbert, the fifth Earl of Carnarvon and sponsor of the expedition, ignored his palmist's warnings and visited the boy king's burial chamber in February 1923. Within two months he was dead – from an infection which developed from a mosquito bite on his cheek.

The story might have ended there, but for a sinister curse engraved above the tomb's doorway which read: 'Death shall come on swift wings to those who disturb the sleep of the pharaoh' – and for several strange events coinciding with the Earl's death.

Not only did the Egyptian capital Cairo suffer a massive power failure, but at his London home the Earl's favourite dog howled and fell dead. Even more intriguingly, when the remains of the boy king were examined, doctors noted a scab-like mark on the left cheek. The blemish was in the same place as the Earl's own fatal bite.

The press of the time latched on to these curious incidents and eagerly awaited news of any further deaths which could be ascribed to what became known as the 'Curse of Tutankhamun'. As it turned out, no less than 22 people who had been involved with the Carnarvon expedition were dead within six years – including 13 men who first worked at the dig. Others, however, survived to a ripe old age, including the excavation's director Howard Carter – an obvious target for the Pharaoh's displeasure.

The Earl of Carnarvon and Howard Carter examine tomb engravings.

THE UNEXPLAINED

water barrel by a jealous lover.

Whether possessions like the Strangler Jacket are receptacles for unknown psychic forces, or simply the focus for their current owners' inner fears remains open to question. Certainly, there is a deadly logic to the fact that the rarer and more valuable the object, the more lurid the stories of the curse surrounding it.

The Hope Diamond, one of many precious stones which have attracted colourful histories, supposedly cost its first owner both his fortune and his life. It was later connected with 18th-century black mass ceremonies, and the doomed queen Marie-Antoinette.

More recently, the stone has been linked with suicide, murder and bankruptcy among its many owners, and is even said to have lost a sultan his throne. Yet in its current home – the Smithsonian Institution in Washington – its keepers have suffered no ill effects. Has the bad luck finally come to an end – or are the stories just another empty curse?

Comic actress Thora Hird on stage in 1973 – was she a victim of the 'Strangler Jacket'.

JAMES DEAN'S CAR

The whole world mourned the death of teen idol James Dean, who was killed in a car accident in 1955. At the time he was driving his brand-new Porsche, despite the entreaties of his friend, the actor Alec Guiness, who had taken one horrified look at the vehicle and implored Dean never to drive it. Was the car cursed? Certainly, events following Dean's death stretch the boundaries of coincidence to the limit.

Car enthusiast George Barris was the first to suffer the so-called jinx when his new purchase slipped from a tow-truck, breaking a mechanic's leg. Shortly afterwards he sold the engine to an amateur racing enthusiast who almost died when his car spun out of control during a competition. And another racing driver was actually killed when a pair of tyres previously fitted to the jinxed Porsche suffered simultaneous blow-outs.

Meanwhile, the damaged body of the car was taken on an ill-fated tour of California to promote road safety. First it slipped from its mountings in Sacramento, injuring a teenager's hip, then a man was killed during a pile-up which occurred while the car was being loaded on to the tour truck. The same truck later suffered brake failure and slammed into a shopfront while carrying the vehicle. Dean's car eventually broke into 11 pieces while on display, before mysteriously vanishing for ever on a train journey back to Los Angeles.

Teen idol James Dean and his ill-fated Porsche roadster.

UNSOLVED MYSTERIES

SHIPPED THROUGH TIME AND SPACE?
CASEBOOK

Nearly 50 years ago, an American ship curiously vanished then reappeared. What really happened?

An eerie greenish fog enveloped a brand-new American destroyer, the *USS Eldridge*, while it was berthed in a high-security dock at the Philadelphia Navy Yard in October 1943. Suddenly, the ship vanished into thin air, only to appear in Norfolk, Virginia more than 350km (approximately 200 miles) away before returning – all in a matter of moments.

Some of the crew, caught in what was later described as a 'force field that surrounded the ship', became semi-transparent; others could be seen distinctly just before the ship was rendered invisible but appeared to be treading on thin air. A few apparently walked through walls, caught fire or were paralyzed before they blurred in shape or disappeared. Some returned to

173

speak of parallel worlds while others were never seen again...

Although this bizarre tale sounds like pure science fiction, it forms the basis of a mystery which has remained unsolved for nearly 50 years. Is there any truth in the story of the Philadelphia Experiment – an allegedly botched US government war-time test that may have transported a ship through time and space?

Basically, the evidence for the whole story revolves around several peculiar letters received by American astrophysicist and ufologist Dr Morris Ketchum Jessup, whose book *The Case for the UFO* was published in 1955. The first, received in October of that year, was signed by a reader called 'Carlos Miguel Allende'; the second, by 'Carl M. Allen', arrived a year later.

Both bore a Pennsylvanian postmark and were written in a rambling, scrawly hand in several different types and colours of pen. In the second letter, Allende recounted the bizarre story of a government experiment he had witnessed which had suddenly gone gravely wrong.

The Experiment Revealed

Allende's account described a US government experiment to render a ship invisible with a force field based on the principles of Nobel Prize-winning physicist Albert Einstein's Unified Field Theory. This endeavoured to explain that electromagnetic, gravitational and nuclear forces were inter-related. According to Allende, the experiment had been a complete success and the ship had vanished briefly into thin air, but the dramatic after-effects had caused the project to be cancelled soon afterwards.

From the safe vantage point of a passing merchant navy ship on which he was serving in October 1943, Allende claimed to have seen the *USS Eldridge*'s crew members suffering the force field's severe repercussions. These included spontaneous combustion, temporary or total disappearance, and sadly, in some cases, madness.

Allende's letter stated that during the experiment the ship was surrounded by a visible globe-shaped field of energy. Before the *Eldridge* vanished, it was apparently enveloped in fog similar to that sometimes seen off the coast of Florida in the area known as the Bermuda Triangle where, to this day, many ships and planes mysteriously disappear. When the *Eldridge* faded from view, the imprint of its hull in the water was clearly visible. Allende also noted that he had seen a newspaper report in a local Philadelphia newspaper which supposedly detailed the sailors' activities after their harrowing experience. At a local bar, the Seaman's Lounge,

WHO WAS CARLOS ALLENDE?

The key to the mystery of the Philadelphia Experiment lies with the mysterious Carlos Allende. Dr Jessup's strange correspondent was known to have had at least five aliases, and was constantly on the move, so it was almost impossible to pin him down.

It was William Moore who finally located the mystery man in the Los Altos region of south-central Mexico. Moore found a tall, middle-aged man with wild eyes who was apt to go rambling on about any number of topics. If finding Allende was a step in the right direction, getting anything out of him concerning the Philadelphia Experiment was another matter. Allende seemed reluctant to discuss the matter and when pressed, changed the subject. He would also make appointments and fail to show up, or appear unannounced.

But circumstantial evidence has backed up some of the claims in his original letters – there was indeed a Carl M. Allen on board a merchant ship in the Philadelphia bay in October 1943 who saw something very strange indeed. 'I actually shoved my hand, up to the elbow, into this unique force field and the flow was strong enough to knock me completely off balance,' he told Moore.

Mystery man Carlos Allende – alias Carl M. Allen – who claimed to have seen the destroyer vanish and then reappear.

UNSOLVED MYSTERIES

The USS Eldridge: was it transported from its berth in Philadelphia (inset top) to Norfolk, Virginia? (inset below)

some of the crew apparently vanished and reappeared in front of bartenders and locals while giving an account of the day's activities. They so distressed the customers that local police were called in to eject them!

Fact or Fiction?

Jessup's first reaction was that the letter was either a hoax or simply the ravings of a disturbed mind. But when the ufologist received a third letter, five months later, which repeated the claims in greater detail, he decided that he would look into the unlikely tale. Unfortunately, he died a few months later.

Word of the Philadelphia Experiment was beginning to spread, and a young college lecturer named William Moore was the next to investigate the mystery. He quickly found his way blocked by official silence, but he persisted and noticed that many of the assertions in the Allende letters appeared to tally with the facts. Moore discovered that there was indeed a USS ship in the Philadelphia docks at the time that the experiment allegedly took place. Furthermore, he dug up material which proved that a merchant ship had been in the vicinity in October 1943, with a certain C. M. Allen among its crew. His increasing belief in the story was further heightened when several other supposed eye-witnesses came forward to back up Allende's claims.

Although their identities have never been revealed, Moore allegedly met and interviewed several research scientists who admitted that they were involved in an 'invisibility' project in the 1940s. The experiment had apparently been conceived as a possible means of protecting ocean-going vessels from enemy fire, and had a high priority rating. Moore also unearthed evidence suggesting that Albert Einstein had been involved in the project since its instigation.

Extra Evidence

Although the mystery of the Philadelphia Experiment looks set to remain unsolved for ever, a similar tale originating from the Soviet Union perhaps gives it greater credence. According to information given to writer Robert Charroux, the authorities believed that the Americans were experimenting with a powerful magnetic field. However, according to the Soviet Union's version of the story, it was a submarine – not a destroyer – which disappeared in Philadelphia and reappeared in Norfolk.

Did American scientists really discover the secret of teleportation, transporting a ship across time and space and unlocking the gates to other dimensions in the process? Sadly, the answer may never be known, for deep in United States government vaults lie a number of Top Secret files which are unlikely to be released for public consumption.

Not that this stops investigators and members of the public from trying to find out. Regular enquiries about the experiment are treated with cordial disdain – they are answered with only a printed regulation denial letter.

THE EINSTEIN CONNECTION

Like most of the evidence related to the Philadelphia Experiment, the assertion that Albert Einstein was involved with it came from Carlos Allende. But investigations have shown that the allegations could have some basis in fact.

Although Einstein died before his Unified Field Theory officially reached fruition, many believe it was already complete. The physicist admitted as much in letters to his friend, the writer and philosopher Bertrand Russell. Indeed, a version of the theory was published in a German scientific journal in 1927.

The theory did not resurface until 1940 when Einstein was apparently convinced that the horror of the German war machine had to be stopped at any cost. He was actually employed by the US Navy in Washington at the time the Philadelphia Experiment was supposed to have happened, and is said to have made a mysterious visit to the ship docks.

Albert Einstein: linked with the alleged US Navy experiment.

UNSOLVED MYSTERIES

MAGICAL DESIGN

Did public taste help to shape the great buildings of the world – or were more mysterious forces at work?

Looking at the high-rise cities of the 20th century, it is hard to imagine architecture having any spiritual or supernatural connection. However, people's first attempts at building were to decorate the sacred sites where they paid homage to the very earliest spirits of nature. When they later began building skywards, it was with as much an eye to the gods as to the demands of commerce. In some parts of the world, at least, it seems that little has changed.

Earth Magic
Standing stones – the mysterious monuments found all over Celtic Britain and France – reflect the strong bond between early architects and the world around them. It is not coincidence that many of the sites have since been found to lie above underground water sources – one of prehistoric man·s most precious commodities. Some researchers also believe that the stones mark 'Earth Energy Points' where the mysterious forces housed within the earth's crust interact in a particularly harmonious way.

Stonehenge, the most famous of all standing stone clusters, is an engineering miracle which took over 500 years to complete. Built in three stages on Salisbury Plain, Wiltshire with the first stage starting around 1900 BC, the stones demonstrate a remarkable knowledge of the universe.

In 1963, Gerald Hawkins, an American astronomer, discovered that many of the stones align with planets and constellations. This

177

THE UNEXPLAINED

spawned the theory that Stonehenge was in fact some kind of primitive astronomical computer which allowed its users to calculate crop planting times and weather cycles. Certainly, the orientation of certain stones is so precise that at the midsummer solstice, the rising sun shines eerily through the gaps between them and falls on what is thought to have once been the altar stone.

Heavenly Guidance

The Great Pyramid of Giza is another breathtaking construction that appears to be mysteriously in tune with the forces of nature. Built around 26 BC on the banks of the Nile as a tomb for the Egyptian Pharaoh Cheops, it is 42 storeys high and covers five hectares (13 acres) of desert.

The scale and orientation of the building allow the sun to cast a shadow on a special floor inside the burial chamber which can be used to calculate the exact time and date. Astronomers have also found that the tip of the pyramid (apex) once aligned with the Egyptian Pole Star – a mystery in itself, as subsequent shifts in the earth's axis mean that the apex is now in alignment with the Western Pole Star.

Lessons learnt from nature also aided the pyramid's surveyors and builders in their mammoth task. The human body itself formed the basis of an ingenious system of measures, while the discovery that water always finds its own level enabled a perfectly flat base to be carved out of desert rock. The pyramid's corners align with the four points of the compass – a feat that could only have been achieved using the stars as a guide – and each is a perfect right-angle.

The Great Pyramid remains an enigma. Like many other ancient sacred buildings, its geometry is based on a magical number square - the square of Mercury. Pyramidologists have suggested that the monolith may hide the secrets of a long-lost but highly developed system of mathematics. Another theory runs that the measurements form a cipher which foretells the future of Christendom up to and including the second coming of Christ. Certainly, the building seems to have magical properties. Food stored within its walls remains fresh for weeks, while knives and razors which are kept at its base never become blunt.

Sun Reflections

Much later, during the 12th century, the architects of the Peruvian Incas demonstrated that they, too, had a special affinity with the natural world.

The Sun Temple in Cuzco was designed to a specific formula to honour the sun god from whom the Incas believed they were descended. Its main feature was a dazzling golden wall decorated

Some believe that Stonehenge was one of the first computers.

The Great Pyramid of Giza, a multipurpose tomb which could be used as a calendar, clock and observatory. Its complicated design was based upon the magic number square of Mercury (below).

8	58	59	5	4	62	63	1
49	15	14	52	53	11	10	56
41	23	22	44	45	19	18	48
32	34	35	29	28	38	39	25
40	26	27	37	36	30	31	33
17	47	46	20	21	43	42	24
9	55	54	12	13	51	50	16
64	2	3	61	60	6	7	57

UNSOLVED MYSTERIES

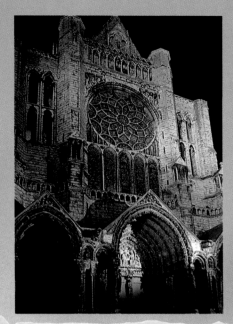

with pictures of the sun, moon, constellations, human figures and a blank centrepiece – the mysterious Creator. The precise orientation of the temple meant that the rising sun lit up the wall in splendour just once a year, marking the beginning of the Incan annual cycle.

Christian architecture too, is not without its mystic symbolism. In the interests of eradicating paganism, many churches were built over ancient sacred sites which themselves marked Earth Energy Points. Some churches also incorporate design features which are found the world over and pre-date Christendom by several centuries – for example, the mysterious spiral maze on the floor of Chartres Cathedral in northern France.

Churches were often built over existing Earth Energy Points.

Wind and Water

In China, architects still obey the rules of their own sacred geometry. Feng Shui (literally, 'wind and water') embodies the belief that people are inextricably linked to their environment by natural forces which make some places feel good to be in, while

THE ART OF GEMATRIA

The Old Testament contains many references to constructions which employed the ancient Jewish art of Gematria – sacred design based on numerological symbolism.

Noah's Ark, for example, is a three-storey structure with each level divided into 11 sections; together they total the sacred Gematric number 33. Animals entered through a door in the lowest level, which represented the plane of physical life. In line with Gematric tradition, the window through which a dove was released to find land symbolized the gap through which the spirit was released to the Creator.

The principles of Gematria were employed by other cultures, including the Babylonians. The Tower of Babel is said to have been constructed to represent a miniature universe, with seven levels corresponding to each of the seven planets known to the ancient world. Its builders climbed seven flights of stairs to the east and descended by seven more to the west, echoing the daily passage of the sun.

Each of the Tower of Babel's layers corresponded to a planet.

THE UNEXPLAINED

LABYRINTHINE LORE

Experts have long debated the significance of the maze – a symbolic design which can be found in art and architecture the world over, cultivated in hedges or built from stone. Some believe it to be a talisman against evil, while others see it as a symbol of mankind's search for the ultimate truth. There is also a theory that the maze signifies the Underworld – a place where the ungodly are trapped and forced to face retribution, while those who follow the 'right path' are released and reborn.

Possibly the most fascinating explanation was put forward in the 1960s by the dowser Guy Underwood, who claimed that the earth's energy forces which he had detected under standing stones formed a pattern resembling seven labyrinthine coils. Traditional mazes often follow a similar pattern; could it be that our ancestors were aware of such forces and sought to copy them in their designs?

Mazes – a quest for truth or a symbol of the Underworld?

others always feel bad.

To this day, a practitioner of *Feng Shui* is consulted before any major construction begins. As well as assisting in the design of the building, he allocates areas to specific uses in a way which is said to bring peace of mind, well-being and good fortune. The Chase Manhattan Bank in Hong Kong, the Citibank in Singapore and the Taiwanese Morgan Guarantee Trust Building are three recent examples of *Feng Shui* in action.

Eternal Designs

On balance, it appears that the forces of nature have had more than a passing role to play in the design and building of many of the world's most remarkable buildings. In Egypt, the Great Pyramid remains standing after more than two thousand years of lashing by desert storms, while in the West, tower blocks and offices face demolition only 20 years after they have been built. Maybe the designs of the ancients hold a lesson for each and every one of us.

Hong Kong ancient and modern: sacred geometry has been used in China for thousands of years.

UNSOLVED MYSTERIES

The search for GOLD

Is it really possible to turn base metals into pure gold?

Win a holiday for two in New York! 'Six chances to win up to £200,000!' 'Enter our Million Pound Bingo!' The possiblities for getting rich quick seem to be omnipresent in our increasingly materialistic society. But such temptations are nothing new, and many people over the centuries have been prepared to risk their lives or, indeed, if the church were to be believed, their souls, to acquire wealth.

But what *is* true wealth? Today, it implies sufficient money on which to retire, buy a second home or perhaps gain personal power. But for primitive cultures, it had an additional aspect as an expression of the

181

THE UNEXPLAINED

BURIED TREASURE

These days, much to the chagrin of the archaeologist, increasing numbers of people set out with metal detectors, intent on finding buried treasure. Others, more in tune with the earth, use forked hazel wands to dowse for metal. It is unlikely, however, that their efforts will turn up the hidden treasures of folk and fairy tales.

Again and again, treasures seem either to have turned to dust or never to have existed in a material sense, like the pot of gold at the end of the rainbow that always seems just beyond that next hill.

Legend has it, for instance, that the last of the Cathars, a Gnostic sect persecuted for heresy by the Catholic French monarchy in the 12th and 13th centuries, succeeded in sending their treasure to safety during the siege of their last stronghold of Montsegur. It has never been found, and many believe that the so-called treasure was not a material one, consisting instead of precious books of Gnostic scripture.

As the old chroniclers constantly remind us, such treasures – often lurking in fairy mounds or in deep caves and guarded by ill-tempered dwarfs or fierce dragons – belong to the world of magic, and the treasure itself may bring ill luck. The theft of the Rhine Maiden's gold in Germanic myth, for example, led inexorably to the doom of Valhalla and the destruction of the gods.

▶ Captain Kidd, one of the most villainous pirates in history, was said to have buried his plundered treasure somewhere on Oak Island, Nova Scotia. To this day, people are still searching for the so-called 'Money Pit', many spending thousands of pounds excavating the site and others losing their lives in the attempt.

divine dimension of existence, and was either to be given to the gods or shared with the community.

It was thus customary for a rich member of an Indian tribe such as the Haida from the north-west coast of America to use his excess wealth to finance a *potlatch* or great feast. Material wealth was thereby transmuted into prestige for the giver: the less wealthy had no opportunity to become envious, and so everyone was happy.

In close-knit and superstitious pre-industrial societies, the appearance of sudden wealth was often ascribed to magical practices or to selling the soul to the devil. Such sinister pacts, reputedly written in blood, are the stuff of occult fiction, notably Goethe's *Faust*. They also formed part of the confessions extracted under torture from witches, when presumably the pain endured was so excruciating that they would have admitted to anything.

When the conquistadors arrived in Mexico and Peru, they were amazed at the abundance of objects fashioned from gold and silver. Gold then, as now, was a standard currency in the West. But to these South American cultures, this precious metal had no commercial value. Instead, its yellow colour made it sacred to the sun god, and so it was used for the adornment of temples and idols.

The virtual indestructibility of gold also made it a symbol of immortality. The mummified body of the young Egyptian pharaoh, Tutankhamun, for example, was encased in three gilt coffins, and a magnificent beaten gold funerary mask was placed over his head. Equipped and guided by relevant texts from the Book of the Dead, the immortal spirit of the young god-king had to traverse the perils of the underworld before rejoining his father, the sun god, Amen-Ra, in the heavens.

The concept of gold and immortality also comes together in the alchemical search for the Philosopher's Stone, which was supposed to turn base metals, such as lead, into gold, and

▶ Tutankhamun's gold-laden tomb containing a funerary mask lay undisturbed until 1922, when it was discovered by Howard Carter.

UNSOLVED MYSTERIES

which also held the secret of eternal life.

Perhaps because the claims were so great, texts on alchemy are usually written in a symbolic language, testimony to both the secrecy that surrounded this dangerous art and its inner spiritual meaning.

The alchemical laboratory was equipped with elaborate vessels, crucibles and furnaces of varying sizes for the 'Great Work'. The operation took nine months and was performed according to beneficial astrological influences, which supposedly assured the alchemist of success.

The secret substance was reduced to a fine powder and placed inside the furnace to be heated by the 'magical' fire. The result of this was known as the 'Black Stage', since the heated material began to putrefy.

The next stage, when a myriad of beautiful colours suddenly appeared, was known as the 'Peacock's Tail'. The final result was the production of a tincture,

▶ King Midas shies away from the daughter his touch has just turned into gold in this Arthur Rackham depiction of the famous cautionary myth about greed.

THE MYTH OF ELDORADO

The first written mention of the fabled wealth of Eldorado, which has lured so many adventurers over the centuries to court financial disaster, if not death, appeared in 1541. The legend, which like all good stories grew in the telling, concerned a golden kingdom somewhere in the newly conquered lands of South America, ruled over by a powerful priest-king known as El Dorado – the Golden Man.

This prince was apparently powdered daily with gold dust, and periodically paddled out by raft to a great lake from which he would throw offerings of gold and precious stones to the tribal deity. Later versions also tell of an annual human sacrifice of a gilded man.

The chroniclers were confident that this treasure could be recovered from Lake Guatavita, in Colombia. Since 1562, numerous expeditions have attempted to drain the lake but have recovered little to merit the outlay. Ironically an exquisite 19.5 cm (7.5 inches) long raft complete with miniature attendants and a larger central figure fashioned from gold was discovered on the edge of a nearby lake in 1856. Perhaps the central core of the legend has more than a grain of truth after all.

▼ The legendary Lake Guatavita in Colombia, where the jewels of Eldorado are believed to lie, and the miniature gold raft that was discovered nearby.

THE UNEXPLAINED

the fabled Philosopher's Stone, which took the form of a fine, reddish powder.

The roots of alchemy lie in the primitive past when man first acquired the ability to work metal. Such metallic substances were regarded as gifts of the earth goddess and were believed to have the capacity to grow underground until they eventually matured into gold.

The alchemist, as a transmuter of metals, was often seen as a godlike figure through his apparent magical powers which enabled him to change the world of nature.

The career of an alchemist could be a hazardous one. Some were derided as wizards, although at the other extreme they were sometimes supported by the monarch. In Prague today

▼ *The alchemist had to go through a complicated succession of heatings, often involving mercury and sulphur, the combination of which resulted in the production of the famous Philosopher's Stone.*

CHANTING FOR WEALTH

Much publicity has been attracted recently by practitioners of what has been described as 'Designer Buddhism'. Devotees are reputed to chant for wealth and other material benefits, both individually and in groups. As usual, the story is a little less obvious than the press would have us believe.

Better known as Nichiren Shoshu Buddhism, this particular sect was founded in Japan by Nichiren Daishonin in the 13th century, taking the Buddhist *Lotus Sutra* as a guiding scripture.

The basic teaching is the spirituality of matter and that material goods are not essentially evil. Today, ten million Japanese participate in the teachings which have spread worldwide since the Second World War.

Chanting the phrase '*Nam-myoho-renge-kyo*' (the title of the holy Buddhist *Lotus Sutra* and meaning broadly 'I identify myself with cause and effect') at a certain pitch is a meditative and cleansing technique which is interspersed, if required, with material requests.

Dedicated practice will often change the chanter's perception of the benefit asked for, so that it may be felt at times to be no longer necessary. Desires become the fuel for enlightenment, and participants claim that an increasing sense of value and joy can be found in their immediate life circumstances.

▶ *Nichiren Buddhism was founded in Japan in the 13th century.*

there still exists the picturesque Street of the Alchemists, known as Golden Lane, where the Holy Roman Emperor, Rudolph II, housed his personal team of alchemists during the 17th century.

The question still remains whether alchemy really works. A large nugget of gold, now housed in the British Museum, was apparently made in 1814, in the presence of a doctor and a colonel. Despite modern scepticism, attempts at practical and esoteric alchemy still continue even today.

UNSOLVED MYSTERIES

AN ESOTERIC ENIGMA

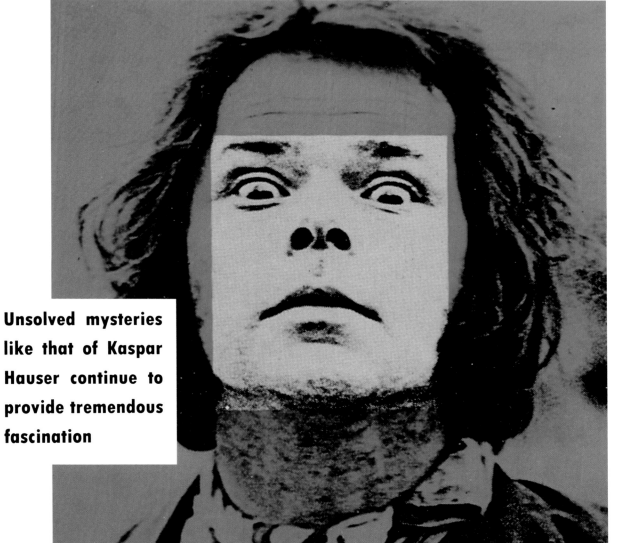

Unsolved mysteries like that of Kaspar Hauser continue to provide tremendous fascination

In May 1828 the German city of Nuremberg was abuzz with a deep and almost obsessive curiosity about a bizarre stranger who had wandered into Unschlitt Square. Who could this shabbily dressed figure possibly be? He was staggering about as if in a daze, was vaguely aristocratic in appearance, yet unable to utter more than a grunt or two. Seemingly, he had a mental age of no more than three, yet although he could hardly even walk, he was physically well into his teens.

Where had he come from? And why was he so distraught? It was almost as if he was a stranger not only to the city of Nuremberg but to the very planet. Rumour and speculation were rife – to such an extent that within days the whimpering figure was to become an object of concern not only to the citizens of this prosperous provincial city but also to the entire population of Germany itself.

Shoemaker George Weichmann had been the first to spot the dishevelled figure, but was unable to elicit much of a response from him. The boy merely muttered what seemed to be his name – Kaspar Hauser – and handed Weichmann a letter, clearly addressed to a certain Captain Wessenig, to whose nearby home the cobbler duly delivered him.

Here, seemingly weak from

THE UNEXPLAINED

hunger, Kaspar was offered a hearty meal, although for some inexplicable reason, he refused it, accepting only plain bread and water which he downed with considerable relish.

Letter from a Stranger

The letter, when opened, proved to be equally enigmatic. Many years previously – so it seemed – its sender had found the child lying on his doorstep and, though a man of limited means, had taken the infant in, keeping him locked in a cellar for all of 13 years, without human contact of any kind. Quite why there had been this need for secrecy was not explained. Circumstances were now such, however, that he could no longer afford to keep the boy and so he had seen fit to deliver him into safe hands. 'If you do not want him,' the letter proclaimed, 'it would be best that you hang him.'

A Bleak Existence

Kaspar had spent his entire life in a dark, confined space, without any sense of time, oblivious to the existence of anyone else, even his jailer. Whenever he awoke, there was always a fresh shirt and bread and water by the bed to which he had been chained; but sometimes the water tasted rather bitter and had the effect of making him sleepy. One day, however, a figure appeared and he was dragged to the centre of Nuremberg where he was given the enigmatic letter and then abandoned.

Public Proclamation

The mayor of Nuremberg, before whom Kaspar was brought, was very moved by his situation and found him a sympathetic soul. It was likely, he announced, that the poor boy had been the victim of some dreadful crime. Anyone with accurate information as to where he had come from would be duly rewarded. Accusations were plentiful, among them the statement that he was surely the illegitimate son of a member of the German royal family. Or had he, some wondered, come from a place way beyond our universe?

A certain Professor Daumer, teacher at a local school, was particularly intrigued by Kaspar; and, indeed, having offered a home to the boy, gradually became utterly captivated by him. Daumer kept a detailed account of the boy's development and character throughout the time he was with him, and was clearly convinced that he must have had supernatural origins of some kind.

Natural Talents

Daumer had noticed that Kaspar, though uneducated, had some remarkable natural talents:

The house where Kaspar Hauser was kept in a dark cellar (centre) before his attempted and eventual murder (below); and a poster and still from the film based on his life (left).

he could identify objects in the dark and was able to locate hidden metal objects, but became intoxicated at the mere sniff of a glass of wine. Under the Professor's tuition he showed an eagerness to learn and a surprising intelligence and was quickly able to produce highly-skilled drawings and writing. Animals, without exception, always responded warmly to him, and even the most savage of dogs lovingly licked his hands.

On 17th October 1829, Daumer's sister noticed traces of blood leading down to their cellar, and there found Kaspar, wounded and unconscious. He was bleeding from his temples, and when he came round he claimed to have been attacked and stabbed by a masked man in a black cloak. Rumour now held that someone must have sent an assassin to get rid of the boy, presumably to prevent revelation of his past history.

Given police protection, Kaspar next went to live with a local lady, Frau Behold, who turned out to be exceedingly cruel when he did not respond to her attempted seduction. Moving on to yet another family, he was now fast catching up at school, though of course in a class of somewhat younger pupils. Socially, however, he remained very much a solitary figure, spending a great deal of time on his own and enjoying what were basically very simple pleasures.

A Mystery Benefactor

Before too long, another letter, coming completely out of the blue, was to cause even greater curiosity, and not only because it enclosed the gift of a diamond ring. It came, apparently, from an anonymous 'friend' who was working on behalf of Kaspar at a distance. When the writer eventually visited, he was found to be a member of the British nobility. There were, he told Kaspar, amazing secrets to be told: but for the moment, Kaspar must simply trust that he had to remain silent until the moment was right. Soon he would return and tell him of his origins, Kaspar was promised. This was not to be, however, for the mysterious benefactor soon perished in the most suspicious and mysterious circumstances.

In 1833, four years after the initial assassination attempt, Kaspar was lured to a Nuremberg park by a stranger's promise that he would be taken to his mother. Here, he was attacked and brutally murdered. No one had witnessed the attack, nor could any weapon be found. But strangest of all, only Kaspar's footsteps could be seen in the December snow afterwards. So it transpired that Kaspar's secrets, of which he was to remain ignorant, were buried with him.

THE UNEXPLAINED

WITCHES' COVENS

For centuries, witchcraft was outlawed, and its followers persecuted throughout Europe. In 1951, however, the British witchcraft laws were repealed, since which time there has been a tremendous resurgence of interest in the pre-Christian mystery religion of Wicca (Saxon for 'life'), which is part and parcel of the practice of witchcraft.

But while the ceremonies and rituals of witchcraft no longer have to remain secret for fear of discovery, some covens do prefer to remain silent about the precise nature of their practices. Secrecy helps to bind the members together, they believe, and also means a much smaller risk of misunderstanding by 'outsiders'.

Secrecy has, not unexpectedly, led to a number of popular

The secret rituals of witches inspire both fear and fascination.

misconceptions about the rituals of witchcraft. Not all covens, for instance, work naked, and followers are adamant that they do not worship the devil, nor are they anti-Christian. Although Satanist cults do exist, they have absolutely nothing to do with the Wicca religion.

SECRET NAMES

The ancient Egyptians are known to have been given two names at birth — one by which they would be called and one which was to remain a secret.

Names, it seems, are regarded as a highly important part of the self to be treated with the utmost respect by many of the world's peoples. Indeed, some will never speak the name of a newborn child for fear that evil spirits could be listening and pick up clues as to the infant's identity.

Likewise, certain Australian aborigines believe it is essential not to disclose what you are called lest you fall victim to unfavourable magical influences. Some of the inhabitants of Papua New Guinea take this even further and do not permit the names of close relatives to be mentioned.

Secret names form part of the culture of the people of Papua New Guinea, and of other societies.

INDEX

Abduction alien 17–20
Abominable Snowman 102, 104
Aborigines 47, 59, 106, 188
Abuse sexual 128
Abydos, Egypt 144
Acupuncture 136
Adams, Sigmund 64
Adolescence 47, 106, 142
Africa
    antelopes 116
    childbirth 143
    masks 127
    plant intelligence 114–15
    shape-shifting 106–7
Afterbirth 143
Agannis, Harry 41–2
Agapoa, Tony 80
Agreda, Mary 31–2
Air 126, 135
Airships 161–4
Akhenaten 90, 91
Alabama, USA UFO sightings 10
Alchemy 182–4
Aliens *see also* extra-terrestrials 9–11, 17–20
Allen, Carl M. 174, 176
Allende, Carlos Miguel 174, 175, 176
Alpha waves 130
Amen-Ra 166, 182
Amulets shamanism 134
Ancestors worship 47
Angel, Jack 121
Angels 141
Animals
    ESP 129
    exorcism 147
    ghosts 103
    healers 104
    totems 106
Antelopes 115, 116
Anthropologists 108
Apollo 85, 87, 108
Apparitions
    hallucinations 57
    ghosts 57
Apports seances 66
Archaeology 73–6, 182
Architecture 177–80
Arctic foxes 100
Arigo 77–8
Arimathea, Joseph of 92
Arion 110
Aristotle 170
Arizona, USA 10, 15
Art 85–8, 156
Arthur, King 91, 92
Astral travel 32
Asthma 135
Astrology 126, 144
Astronomy
    Great Pyramid of Giza 178
    Stonehenge 177–8
Asvins, Hindu gods 140
Atherstone, Lieutenant Commander 164
Atlantis 76, 169–72
Auras 135, 156
Australia
    Hanging Rock 13–14
    secret names 188
    shamans 79
    totem animals 106
    Yarralumba House ghost 59
Autistic children 86
Automatic writing 74, 83, 88, 143
Automatism 88
Avatars Hinduism 153, 155
Avebury, Wiltshire crop circles 27
Azande peoples 143

Babel Tower of 179
Babylonians 179
Bach, Johann Sebastian 114
Bailey, Ethel 79
Ball lightning 99, 124, 160
Baptists 149
Baranowski, Frank 156
Bards inspiration 85–6
Barr, Gunther 71
Barris, George 168
Bartlett, John 73–4
Bathory, Elizabeth 46
Bathurst, Benjamin 16

Bavaria 64, 150
Beardsley, Aubrey 143
Beardsley, Richard 38
Beasts
    Exmoor 97–9
    Gèvaudan 100
Beauterne, Antoine de 100
Beauty and the Beast 108
Beauvais, France 161, 164
Beer, Trevor 98
Beere, Richard 73–4
Beethoven, Ludwig van 44, 83
Behold, Frau 187
Belgium stigmata 150
Belief ESP 132
Belladonna 106
Bender, Hans 64, 71
Benin masks 127
Beraud, Martha 67
Berkeley Square, London 59–60
Berlitz, Charles 159
Bermuda Triangle 157–60, 174
Betting dreams 35
Bettiscombe Manor, Dorset 59
Bible 144
Big Hairy Monsters (BHMs) 102
Bigfoot 103
Bile 135
Bimini 76, 171
Birmingham, England 120
Birth cycles 144
Black Stage 183
Blackburn, William 94
Blake, William 85, 87, 141–2
Blanks, Edward 98
Blavatsky, Madame 172
Bleak House 122
Bligh Bond, Frederick 73–4
Blood 104, 135
Blythburgh Black Dog 99
Boleyn, Anne 50
Bonner, Gilbert 71, 72
Book of the Dead 182
Booth, David 68
Borley Rectory, Essex 58
Bosch, Hieronymus 128
Brahms, Johannes 114
Brazil surgery 77–9, 151
Bridges black dogs 99
British Museum 144, 166, 184
Brontë sisters 127
Brown, Rosemary 44, 83
Buddhism Nichiren 184
Bull, Alex 80
Bungay, Suffolk 99
Burma soul transference 108
Bush souls 107

Caddy, Eileen 116
Caddy, Peter 116
Calabar peoples 107
Calculation autism 86
Caldwell, Taylor 115
California, USA
    Bigfoot 103
    channelling 127
    Racetrack Playa 117–18
    stigmata 149–50
    UFO sightings 11, 22
Campbell, Virginia 142
Canada 104, 114
Canadian Archaeological Association 75
Canberra, Australia 59
Cancer
    animal healers 104
    Kirlian photography 135
Cardiac, Jean 144
Carnarvon, Earl of 167
Carter, Eliza 16
Carter, Howard 182
Carter, Jimmy 11
Cass, Raymond 71
Castor 140
Cat Protection League 104
Cathars 182
Cats 97–100, 104
Cayce, Edgar 76, 136, 170–1
Celts 85, 90
Central Intelligence Agency (CIA) 22
Cereology 26–8
Channelling mediums 84, 127
Chantilly, France 60

Chaplin twins 137–8
Chapman, George 79
Charles I 58, 136
Charroux, Robert 176
Chartres Cathedral 179
Chastel, Jean 100
Chess child prodigies 88
Childbirth 143
Children
    autistic 86
    ESP 141–4
    prodigies 88
China
    ancestor worship 47
    Feng Shui 179–80
    geomancy 55
    telepathy 130
    wild men of Hupeh 101–2
Choleric humour 126
Christianity
    architecture 179
    faith healing 136
    Holy Spirit 87
    inspiration 86
Christides, Nick 110
Christo, Fabianoda 78
Church
    black dogs 99
    exorcism 145–8
    stigmata 149–51
Churchill, Winston 24
Churchwood, James 170
Cities phantom 171
Clairvoyance 83–4, 156, 172
Clothing
    aliens 9
    Strangler Jacket 167
Cobbett, William 98
Collins, Andy 89–92
Collins, Joan 42
Colombia Eldorado 183
Colossus Memnon 119
Colours
    auras 156
    healing 136
    scrying 131
    vibrations 132
Combustion spontaneous 121–4
Communication aliens 10
Computers hauntings 61–3
Conan Doyle, Arthur 34, 68, 159, 162
Condon Committee 24
Conquistadors 182
Cook, Florence 93–6
Corder, William 166, 167
Corinth, Greece 110
Coronation services 96
Cosmology 32
Coster, Ian 162
Cottingley Fairies 34
Covens 188
Cowan, Elsie 154
Cowan, Walter 154
Crandon, Margery 67
Creatures mythical 101–4
Crete 171, 172
Crocodiles 100
Croiset, Gerard 160
Crominski, Joyce 139
Crookes, William 94–5
Crop circles 25–8
Crowly, Tom 102
Cruz, Sonja da Costa 77–8
Crystal balls 131
Cuffley Lioness 98
Curran, Pearl 88
Curses 46, 165–8
Cutajar, Andrew 39
Cuzco Sun Temple of 178–9
Cyclops 158, 159

Daishonin, Nichiren 184
Daniken, Eric Von 76
Daphne 108
Darbishire, Stephen 23
Dartmoor cat 100
Daumer, Professor 186–7
Davis, Jim 42–3
Dean, James 168
Death, moment of 51–2
Dee, John 92
Dei Gratia 159

Delhi, India 144
Demons exorcism 145
Depression 113
Detroit, USA 123
Devas 116
Devi, Shanti 144
Devil 26, 127, 147, 182
Devon porcupines 100
Dickens, Charles 87, 122
Diet extra-sensory perception 131
Disappearances 13–16, 37–8
Dissociation 127
Dobbs, George 50
Dogon peoples mythology 30
Dogs 99
Dolphins 109–12
Donnelly, Ignatius 172
Dorset screaming skull 59
Dostoevsky, Fyodor 70
Downing Street, London 59
Dowsing
    buried treasure 182
    extra-sensory perception 84
    labyrinths 180
    psychic archaeology 74
    standing stones 27
Dr Jeckyll and Mr Hyde 126
Dracula 46, 96
Dragon Project 118–19
Drawing autistic children 86
Dreams
    astral travel 32
    betting 35
    ghosts 57
    parallel lives 15
    psychic quests 89
    reincarnation 127
Drink abstention from 16
Drugs hallucinogenic 106
Duncan, Helen 68
Dunglas Home, Daniel *see* Home ...
Dunne, J. W. 15
Dunstable, Bedfordshire 37–8
Dunstall Castle 90
Durer, Albrecht 143
Dyer, George 24

Eady, Dorothy 143–4
Earth 126, 135
    Energy Points 177, 179
    Goddess 184
Easter Island statues 12
Ectoplasm mediums 66, 68
Edison, Thomas 70
Edmonds, John 34
Egypt
    architecture 178, 180
    astral travel 32
    Atlantis 169
    curses 165–7
    gold 182
    healing 134
    psychic quests 90, 92
    secret names 188
    UFO sighting 29
    Virgin Mary 52
Egyptology 166
Einstein, Albert 15, 174, 176
Eisenbud, Professor 36
Eisenhower, Dwight 11, 22
Elder trees properties 115
Eldorado 183
Electricity 48, 62–3
Electromagnetism
    Bermuda Triangle 160
    crop circles 27
    geopathic zones 28
    Kirlian photography 135
    Unified Field Theory 174
Electronic Voice Phenomena (EVP) 70–2
Elements 135
Eliot, George 87
Elizabeth I 58
Elizabeth II 58, 96
Ellis, David 70–1
Emerson, Norman 75, 76
Emotions 113–14, 127, 142
Empathy twins 137–42
Empedocles 135
Enchantment 106
Energy stones 119

Enfield poltergeist 45–6
Enlightenment Buddhism 184
Epilepsy 147
Eroticism 95, 96
Essex witch of Scrapfaggot Green 118
Ethics dolphins 109–10, 112
Europa 107
Evil Eye 143, 165
Evolution ESP 129
Exmoor Beast of 97–9
Exorcism 145–8
Extra-sensory perception (ESP) 81–4, 129–32, 141–4
Extra-terrestrials encounters with 9–11, 159
Extra-terrestrials see also aliens
Ezekiel 9, 141

Faces reading 128
Faerie-Wildfolk 99
Fairies
    Cottingley 34
    Ireland 16
    kidnappings 20
    rings 115
    visions 141
Faith healing 136
Fakirs 150
Familiars 106, 107
Family
    curses 46
    gods 48
Farnham, Surrey 98
Faust 182
Fawcett, Percy Harrison 171–2
Feminism 95
Feng Shui 55, 179–80
Fetishism 95
Fields, Totie 42
Findhorn, Scotland 116
Fire 126, 135
Fishermen 108
Flannan Islands Lighthouse 158
Florida, USA 112, 122–3
Flowers 66, 113–16
Flying cats 104
Flying saucers New Mexico 21–2
Flying saucers see also unidentified flying objects
Fodor, Nandor 115
Folklore Ireland 16
Food 16, 131
Forfar, Archie 158
Fort, Charles Hoy 7–8, 16
Fortune-telling 128
Fox, Kate 67
Fox, Margaret 67
Foyster, Marianne 58
France
    Beast of Gèvaudan 100
    Cathars 182
    Chartres Cathedral 179
    exorcism 147
    ghost 60
    plant intelligence 114
    R101 crash 161, 162
    reincarnation 127
    UFO reports 11, 12, 24
Freemasons 91, 92
Frei, Gerhard 72
Freitas, Jose Pedro de (Arigo) 77–8
Fritz, Adolpho 78
Fulton, Roy 37–8

Gaea 108
Galgani, Gemma 150
Ganzfeld technique 132
Garrett, Eileen 76, 161–4
Geddes, Barbara Bel 42–3
Geiger counters 119
Geller, Uri 36, 82, 143
Gematria 179
Genius 87–8
Geomagnetism 124
Geomancy China 55
Geometry Pyramids 178
Geopathic zones 28
George I 136
George III 58
Geraniums sensitivity 114
Germany
    ghosts 63, 64
    Kaspar Hauser 185
    Rhine Maiden's gold 182
Gernon, Bruce Jr 159

Gèvaudan, France 100
Ghosts
    animal 103
    commercial premises 53–6
    examples 49–52
    hitch-hikers 38
    houses 57–60
    lucky 44
    show business 41–4
    work 61–4
Giza Great Pyramid 178, 180
Glamis Castle ghosts 58
Glastonbury, Somerset 73–4, 84, 92
Gleason, Jackie 22
Gnomes 20
Gnostics 182
Godfrey, Alan 10
Gods family 48
Goethe, Johann Wolfgang von 182
Gogh, Vincent Van 83
Gold 181–4
Good, Timothy 22, 24
Goodenough, Florence 144
Goodman, David 98
Gool, Christopher 139
Goya, Francisco de 88, 143
Grabandal, Spain 52
Grail Holy 92
Grand Bahama Island 157, 158
Grand National 88
Grandier, Urban 148
Grange, Anton Le 39
Gravity 30, 31
Great Isaacs Lighthouse 158
Great Leighs, Essex 118
Great Pyramid of Giza 178, 180
Greece
    ancient 85, 109–10, 140, 169
    childbirth 143
    werewolves 106
Green, Yvonne 139
Green-fingers 116
Greenhaus, Jeff 10
Greenpeace International 109
Gremlins 61–4
Grosse, Maurice 46
Ground Saucer Watch (GSW) 22
Guatavita, Lake Colombia 183
Gudkov, Nikolai 61
Guinness, Alec 168
Gunderson, Anne 158
Gunpowder Plot 90
Guppy, Mrs 31, 66, 94

Haida peoples 182
Hailstones 120
Hallucinations
    celestial cities 171
    ghosts 49–50, 141
    paranormal phenomena 29
    werewolves 105
Hallucinogenic drugs
    masks 127
    shamanism 134
    werewolves 106
Hamelin, Germany Pied Piper 147
Hampshire crop circles 25, 28
Hampton Court Palace 58
Hands laying on of 136, 151
Hanging Rock picnic at 13–14
Hankey, Rosemary 38
Hannah House, USA 60
Haraldsson, Erlendur 155–6
Harden, Tomas 63
Harrison, Bridget 138
Harvey, Laurence 42
Hauntings see ghosts
Hauser, Kaspar 185–7
Hawaii 108, 119–20
Hawkins, Gerald 177–8
Healing
    animals 104
    holistic 133–6
    plants 115
Heavy Metal music 114
Heiner, Friedrich 144
Henry VIII 50, 58
Herbert, George 167
Heresy 182
Herman, Louis 110
Herodotus 110
Hertfordshire ghosts 53–6
Hexham, Northumberland 139
Hill, Barney 10
Hill, Betty 10

Himalayas 102–3, 170
Himmler, Heinrich 69
Hinchcliffe, Emilie 163
Hinchcliffe, Walter 163
Hinduism
    avatars 153, 155
    fakirs 150
    twin mythology 140
Hippocrates 135
Hippotami 107
Hird, Thora 167–8
Hirst, Bertha 164
Hitch-hikers phantom 37–40
Holistic medicine 133–6
Holm, Peter 42
Holy Grail 92
Holy relics 89
Holy Spirit 86, 87
Holyrood 136
Home, Daniel Dunglas 67, 94–5, 105
Homeopathy 136
Hong Kong Feng Shui 55, 180
Honorton, Charles 132
Hope Diamond 168
Hopkins, Robert Thurston 166–7
Hormones birth cycles 144
Horse chestnuts 115
Horse racing 35
Horton, Edith 13–14
Hoskins, Bob 44
Houdini, Harry 68
Houses ghosts 57–60
Hoven, Wouter van 114–15
Howard, Katherine 50, 58
Huffsticker, Carol 76
Humanoids 9–11, 21–2
Humours 126, 135
Hung, Carl 38
Hupeh, China 101–2
Hussey, Ben 13–14
Hynek, J. Allen 24
Hypnosis
    alien abduction 19, 20
    past-life therapy 127
    UFO encounters 10, 31
    werewolves 105
Hysteria
    Devils of Loudon 148
    paranormal phenomena 29
    werewolves 106

Images paranormal 36
Impressions psychometric 132
Incas 178–9
India 127, 140, 143, 150
Indians, American
    ancestor worship 47
    archaeology 75–6
    feasting 182
    totem animals 106
Infant of Lubeck 144
Infra-red photography 96, 119
Initiation rites 106, 134
Insomnia UFO enounters 10
Inspiration art 85–8
Institute of Paranormal Research 64
Institute of Psychiatry 65
Intelligent Life Elsewhere (ILE) 32
Intuition 127, 156
Invernesshire 99
Invisibility 174, 176
Ipswich, Suffolk 122
Ireland 16, 107, 108, 110
Irwin, 'Bird' 162, 163–4
Isis Temple of 144
Islam 150, 151
Italy 150, 151

Jaarsveld, Dawie van 39
Jainar, Marcelo 80
James I 90
Janssen, David 42
Japan 47, 108, 184
Jessup, Morris Ketchum 174–5
Jesus Christ 92, 96, 146, 151
Jinxes 42, 166, 167, 168
Joan of Arc 89
Johnston, Squadron Leader 163, 164
Joseph of Arimathea 92
Ju-ju men 107
Judas Iscariot 96
Jung, Carl 30–1, 70
Jupiter 140
Jurgenson, Friedrich 69

Kalki 155
Kardec, Allan 79
Kawasaki Heavy Industries 61
Kennedy, John F. 70
Kensington, London 63
Kenya 120
Kersten, Felix 69
Kidd, Captain 182
Kidnappings medieval 20
Kilner, Dr 166–7
King, Katie 93–6
King's Evil 136
Kirlian photography 135
Kissing 96, 108
Klimt, Gustav 96
Knights Templars 90, 91
Knossos, Crete 171
Krishna 155, 156
Krogman, Wilton 123
Krishna 155
Kurma 155
Ku 108
Kwakiutl peoples 143

Labinkir, Siberia 101
Labyrinths symbolism 180
Lake Guatavita, Colombia 183
Landa twins 139
Lang, William 79
Lares Roman gods 48
Larkin, Philip 71
Last Supper 92
Lateau, Louise 150
League of Nations 35
Leda 140
Leeches 135
Leeds Poltergeist 147
Lemuria 172
Lennon, John 44, 68
Lennon, Julian 44
Leonard, Gladys Osbourne 67
Leopold, Irma 14
Lethbridge, Tom 74–5
Levitation
    alien abduction 20
    avatars 154
    Eusaphia Palladino 35, 67–8
    St Theresa of Avila 30
Lewis twins 138
Ley, Eric 98
Ley lines 27, 103
Lie Hong-wu 130
Light, Gerald 22
Lilies 115
Lilly, John 111–12
Lincoln, Abraham 34, 58, 68
Lincolnshire ghosts 103
Liszt, Franz 44, 83
Llanidoes, Wales 39
Loch Ness monster 24, 101, 102
Lockhart, June 156
Loffert, Ralph 119
London, England
    ghost 59
    phantom bus 63
    Spiritualist Alliance 163
    West Ham disappearances 16
Los Angeles, USA 42
Lotus Sutra 184
Loudon, France Devils of 148
Louis XV 100
Lourdes, France 32, 136
Lowe, Dorothy 138
Lubeck, Germany 144
Lutsch, Hans 71
Lycanthropy 105–6
Lyttleton, Edith 35

Macanas, Felisa 80
Macaulay, Thomas 140
McCraw, Greta 13–14
Machines ghosts 61–4
McKusick, Marshall 76
Maclean, Dorothy 116
McMullen, George 75
Magic
    familiars 107
    masks 127
    medicine 133–6
    plants 115
Maimonides Medical Center 132
Malzieu 100
Manet, Edouard 83
Manning, Matthew 84, 88, 143
Maoris inspiration 86
Marconi, Guglielmo 70

Marie-Antoinette, Queen 168
Marigolds 114
Marks reincarnation 127
Martindale, Harry 62
Mary Celeste 157, 159
Mary, Queen of Scots 90, 91
Masks 127, 182
Masser, Ellen 132
Mathematics 88, 178
Matsaya 155
Mauna Loa volcano, Hawaii 120
Mayans 170
Mayfair, London 59
Mazes 179, 180
Medicine 133–6
Meditation 130, 151, 184
Mediums
  channelling 84, 127
  children 142
  ectoplasm 68
  messages 65–9, 82–3
  psychic quests 92
  R101 crash 161–4
  Victorian 93
Medjugorje, Yugoslavia 52
Medusa 126
Meek, George 80
Melancholic humour 126
Memnon, Egypt 119
Memory 86, 127
Men in black (MIB) 23
Menstruation 47, 138
Mercury
  Roman god 55
  square of 178
Merry Maidens standing stones 27
Metal
  alchemy 184
  detectors 182
Meteorology 27
Mexico 182
Michaelis, Inquisitor 147
Micronesia 172
Midas, King 183
Middle Ages 104, 145
Midlands, England psychic quests 89–92
Mimosa psychic abilities 115–16
Mind
  alien abduction 20
  extra-sensory perception 82
  plant sensitivity 113
  psychic surgery 80
Ministry of Defence (MOD) 23, 24
Minoans 110, 171–2
Mirabelli, Carlos 67
Mirages 171
Mistletoe 115
Mohammed 151
Moment-of-death apparitions 51–2
Monarchy British 136
Money Pit 182
Montgomery, William 107–8
Montsegur 182
Moon birth cycles 144
Moore, David 147–8
Moore, Dr 16
Moore, Janet 147–8
Moore, William 174, 175–6
Moray, Scotland 100
Morgan, Henry 94
Morgawr 160
Mormons 47
Morris, Desmond 104
Moses, Stainton 66
Mount St Helens, USA 40
Mozart, Wolfgang Amadeus 83, 88, 144
Mu 170
Multiple personalities 128
Mumler, William 33–4
Murray, Douglas 166
Muses 85, 86, 87, 88
Music 86, 88, 114
Mythical creatures 101–4
Mythology twins 140

Naacal 170
Naddair, Kaledon 99
Names
  secret 188
  werewolves 106
Napier, John 104
National Aeronautics and Space Administration (NASA) 11
Nature psychic abilities 113–16
Near-death experiences 32

Nebraska, USA 12
Necrophilia 96
Neff, William 16
Neil-Smith, Christopher 148
Neumann, Therese 150
New Hampshire, USA 10
New Mexico, USA 21–2
New York State, USA 17–20
New York, USA 16, 41
New Zealand 139–40
Nichiren Buddhism 184
Nichol, Agnes 66
Nightmares 10
Nixon, Richard 22
Noah's Ark 179
Noble, Ted 98–9
Norfolk, England 39, 55–6, 60
North America reincarnation 127
Northampton, England 100
Norwich, England 55–6
Nottingham, England 98
Nova Scotia 182
Noyes, Ralph 24
Nrisinha 155
Nuremberg, Germany 185–7

Oak properties 115
Ohio, USA twins' convention 138
Old Testament 179
Old wives tales shops 56
Old wives tales see also superstitions
Ono, Yoko 68
Orbito, Alex 80
Oregon, USA 40
Orient Express train 63
Ossowiecki, Stefan 75
Ottawa, Canada 104
Ouija boards 88
Out-of-body experiences 32
Ovonnik 108
Oxfordshire, England 27

Padre Pio 150, 151
Pagans mistletoe 115
Painting automatic 83, 143
Palladino, Eusapia 34–5, 36, 67, 68
Papua New Guinea secret names 188
Parallel lives 15, 138
Parapsychology Foundation 76
Parapsychology Society of Spain 139
Parasearch 89–92
Parasurama 155
Pascal, Blaise 144
Past-life therapy 126–7
Peacock's Tail 183
Penates Roman gods 48
Pendulums 27, 84
Penn, Sybil 58
Pentecost 86
Pereora, Maria 36
Personality medieval 126
Peru 182
Peterson, Billy 123
Pett, Grace 122
Pharaohs 165–6, 167, 178, 182
Philadelphia Experiment 174–6
Philippines psychic surgery 79, 80, 151
Phillips, Graham 89–92
Philosopher's Stone 182, 184
Phlegmatic (phlegm) humour 126, 135
Phobias 126–7
Photography 33–6, 135
Physiognomy 128
Picasso, Pablo 83, 88, 143
Picknett, Lynn 48
Pied Piper of Hamelin 147
Pilsdon Pen, Dorset 75
Piper, Leonora 67
Placentas 143
Plants psychic abilities 113–16
Plato 169, 170, 171, 172
Playa stones, USA 117–18
Playfair, Guy Lyon 46
Podbrdo Hill 31
Poetry inspiration 85–6
Pole Stars 178
Pollock twins 139
Pollux 140
Poltergeists
  Borley Rectory 58
  children 142
  examples 31–2, 45–8
  fires 124
  haunted office equipment 64
  Leeds 147

Matthew Manning 88
  Scrapfaggot Green 118
Poniatowski, Stanislaw 75
Pontefract, Yorkshire 47
Pooka 107, 108
Porcupines 100
Poseidon 169, 170
Possession 46, 125–8, 145
Potlach 182
Prague, Czechoslovakia 184
Prashanthi Nilayam 154
Pratt, Gaither 68
Prayer 146, 151
Predictions 129, 131, 163
Prehistoric sites 99, 177
Premonitions 68
Presley, Elvis 44
Prestige 182
Price, Harry 58, 63, 162–4
Prichard, Dianne 47
Prichard, Philip 47
Priests exorcism 146, 148
Prodigies child 88
Projects
  Blue Book 23, 24
  Dragon 27, 118–19
  Grudge 23
  Sign 23
  Twinkle 23
Prophesy 40
Psychiatry 126
Psychic abilities plants 115–16
Psychic photography 33–6
Psychic quests 89–92
Psychoanalysis 105
Psychokinesis 31, 82
Psychology 83
Psychometric impressions 132
Psychometry 75, 84
Psychosomatic illness 135
Psychotherapy 83
Psychotics 105
Puberty 142
Puharich, Andrija 78
Pumas Britain 97–100
Puttaparti 154
Pyramidology 178
Pyramids 166, 178, 180

Quade, Marion 14
Quakers 88
Quests 89–92

R101 crash 161–4
Raccoons 100
Racetrack Playa, USA 117–18
Rackham, Arthur 183
Radhakrishna, V. 154
Radio carbon dating 75
Rama 155
Ramachandra 155
Randles, Jenny 42, 43
Raudive, Konstantin 70, 72
Ravens 143
Raynham Hall, Norfolk 60
Red Barn murder 166
Red Rum 88
Reeser, Mary 122–3
Reincarnation 127, 139, 144, 155
Relativity theories of 15
Relics psychic quests 89
Religion 95
Renaissance 136
Rendhell, Fulvio 96
Repression sexual 95
Rheumatism plant healing 115
Rhine Maiden's gold 182
Richard E. Byrd 158
Rituals
  childbirth 143
  masks 127
  psychic quests 89–92
  shops 55
  totem animals 106
Rivail, Leon 79
Robots industrial 61
Rocks properties 117–20
Rollright Stones, Oxfordshire 27, 118–19
Romans ancient 48, 140
Rome, Italy 124
Roosevelt, Theodore 58
Rosenheim, Germany 64
Roses 115, 116
Round Table 92

Roux, Maria 39
Royal Air Force (RAF) 24
Royal Marines 97–8
Royston, Hertfordshire 56
Rudloe Manor, Wiltshire 24
Rudolph II, Holy Roman Emperor 184
Rudraksha 156
Runcie, Robert 146
Ruppelt, Edward 23
Russell, Bertrand 176
Russia plant intelligence 114
Russia see also Union of Soviet

Sai Baba, Sathya 153–6
St Albans, Hertfordshire 54–5
St Elmo's fire 140
St Francis of Assisi 150–1
St Helens, Mount, USA 40
St John of the Cross 86
St Mary the Virgin 32, 52
St Theresa 30, 86
Salisbury Plain, Wiltshire 177
Salt 143
Sandwich, Kent 100
Sanguine humour 126
Satanists 188
Sauvin, Pierre Paul 113–14
Savalas, Telly 41–2, 43
Scars reincarnation 127
Schizophrenia 126, 127, 147
Schneider, Anne-Marie 64
Schneider, Rudi 66
Science
  astral travel 32
  birth cycles 144
  dolphins 109
  electromagnetism 28
  ESP 132
  exorcism 147
  mythical creatures 104
  paranormal phenomena 29
  Philadelphia Experiment 174–6
Sclater, P. L. 172
Scotland
  black cats 100
  mythical creatures 102
  poltergeists 142
  shamans 99
  soul transference 107–8
Scotland Yard, London 103
Scrapfaggot Green, Essex 118
Screaming skulls 59
Scrofula 136
Scrying 129
Scully, Frank 21–2
Sea serpents 160
Seances
  apports 66
  examples 67
  ghosts 50, 55
  psychic quests 92
  Strangler Jacket 167
  Victorian 93
Seasons birth cycles 144
Second World War 184
Self-hypnosis 105
Sennah, Abderrahman 39
Serios, Ted 35–6
Sexual abuse multiple personalities 128
Sexual repression mediums 95
Seymour, Jane 58
Shamans
  healing 134
  psychic surgery 78–9
  Scotland 99
  shape-shifting 107
  spirit creations 50
Shan, Lawrence Le 76
Shape-shifting 105–8
Sharp, Robert 117–18
Shoplifting 56
Shops 55–6
Show business ghosts 41–4
Showers, Mary 95
Shuckey Dog 99
Siberia familiars 107
Silbury Hill, Wiltshire 27
Silva Mind Control 130
Silver 182
Singapore Feng Shui 180
Sirius star system 30
Sitwell, George 92, 96
Sixth sense 129, 131, 132

Skryker 99
Skulls
    haunted 166
        screaming 59
Skyfalls examples 30–1
Slaughterford, Christopher 59
Slavery ghosts 60
Smells ghosts 60
Smithsonian Institute 168
Smyth, Frank 19
Snowman Abominable 102–3
Social conscience dolphins 109–10
Society for Psychical Research 46, 50–1, 58, 66
Society for the Study of Supernatural Pictures 34
Solon 169
Solstices 171, 178
Somerset, England 39–40, 92
Souls
    familiars 107
    selling to Devil 182
    spirit possession 125
    transference of 107–8
South Africa plant intelligence 115
Soviet Union see Union of Soviet Socialist Republics
Soya beans 114
Spain Parapsychology Society 139
Spells 106, 134
Sphinx 76
Spielberg, Steven 24
Spinelli, Ernesto 141
Spirit creations 50
Spirit guides 66–7, 74, 84
Spirit possession 125–8
Spiritist Society 79
Spiritualist Alliance 163
Spiritualists
    ancestor worship 47
    children 142
    churches 65
    Conan Doyle 162
    ESP 82–3
    ghosts 50
    mediums 93, 127
    psychic photography 34
    R101 crash 163, 164
    teleportation 31
Split personalities 126
Spontaneous combustion 121–4
Spoon-bending 81, 82
Springer, Arthur 103
Squid giant 102
Standing stones 27, 118, 177, 180
Stately homes ghosts 58
Steiner, Rudolf 172
Steinman, William 21–2
Stevenson, R. L. 126
Stigmata 149–51
Stoker, Bram 46, 92, 96
Stokes, Doris 83, 84
Stonehenge, Wiltshire 27, 74–5, 177–8
Stones
    properties 117–20
    standing 27, 118
Storsjoodjuret 101
Strange Phenomena 89

Strangler Jacket 167
Streams black dogs 99
Strieber, Whitley 17–20
Stringfield, Leonard 22
Subconscious automatic writing 83
Sumatra totem animals 106
Sunderland, Gaynor 90
Superstitions 56, 79, 115
Surgery psychic 77–80, 151
Surrey, England 98, 99
Swans psychic quests 91
Sweden Storsjoodjuret 101
Swedenborg, Emanuel 66
Swords psychic quests 90–2
Symbolism
    architecture 179
    labyrinths 180
    totem animals 106

Taboos 106, 165
Taenarum 110
Taiwan 55, 128, 180
Taman peoples 108
Tanous, Alex 68
Tanuki 108
Tape recordings EVP 69–72
Telepathy 81, 112, 129–32, 140, 143
Telephones ghosts 64
Teleportation 16
    avatars 154
    Philadelphia Experiment 176
    poltergeists 46
    seances 66
    spiritualists 31
Television poltergeists 48
Tenerife 102
Tenuto, Bill 44
Thailand 110
Thamyris 86
Thebes 166
Theosophical Society 172
Therapy 104, 126–7
Thomson, Christopher Birdwood 161
Thoughtographs 35–6
Thurley, Len 120
Tibet 50, 89
Tides birth cycles 144
Tigers 108
Tintagel, Cornwall 91
Tomatoes 114
Totem animals 106
Totem poles 106
Trances 65–6, 134, 171
Transcendental meditation 130
Transmutation 181–4
Transylvania 46
Trauma 128
Treasure buried 182
Trees magic 115
Tulpas 50
Turtles 108
Tutankhamun 167, 182
Tutin, Dorothy 43
Twins 137–41

Ufologists Philadelphia Experiment 174, 175
UFOs see unidentified flying objects

Ultrasound Rollright Stones 27, 118–19
Underwood, Guy 180
Underworld 180
Unidentified flying objects (UFOs)
    ancient Egypt 29
    crop circles 28
    dolphins 112
    Hanging Rock 14–15
    Jimmy Carter 29–30
    pattern of encounters 9–11
Unified Field Theory 174, 176
Union of Soviet Socialist Republics (USSR) 101, 176
Uniondale 39
United Kingdom (UK)
    ghosts 37–8, 39–40, 59
    phantom bus 63
    standing stones 27
    UFO reports 23, 24
United States Air Force (USAF) 21, 22–3
United States of America (USA)
    ghosts 38, 40, 41–3, 58
    Philadelphia Experiment 174
    UFO sightings 10–12, 15–16, 17–20, 21–4
Unsworth, Harold 39–40
Urada, Kenji 61
Urquhart Castle 102
US Navy Bermuda Triangle 157, 158
USS Eldridge 173, 174, 175
USS Tigrone 158

Valhalla 182
Vallee, Jacques 24
Vamana 155
Vampires 46, 92, 96
Van Gogh, Vincent 83
Varaha 155
Vardy, Edgar 66
Vatican 151
Venice Festival 127
Venus flytrap 115
Vesta 48
Vibhuti 154
Vibrations 113, 132
Villa, Paul 12
Villiers, Oliver 163
Vishnu 155
Visions 89, 141
Volcanoes 40, 120
Volckman, Mr 93, 94

Wales 100
Wallace, Alfred 34
Walpole, Dorothy 60
Walton, Travis 10, 12, 15–16
Warboys, Robert 59
Warner, W. Lloyd 79
Warts plant healing 115
Washington, USA 40
Watson, Lyall 62–3, 114, 119
Water 126, 135, 177
Waverley Abbey, Surrey 98
Wealth 181–2, 184
Weather
    Bermuda Triangle 159–60
    electromagnetism 28
    Stonehenge 178

Webster, Ken 63
Weichmann, George 185
Wells faith healing 136
Werewolves 104, 105–6
Wessenig, Captain 185
West Ham disappearances 16
Westbury crop circle 26
White Brotherhood 44
White House, USA 58
Wicca 188
Willidjungo 79
Willows plant intelligence 114
Wiltshire
    crop circles 25
    Stonehenge 177
    UFO research 24
Wiltshire, Stephen 86
Wimble Toot, Somerset 92
Windsor Castle ghosts 58
Witch doctors
    healing 134
    shape-shifting 107
Witchcraft
    confessions 182
    covens 188
    exorcism 146–7
    Ireland 16
    masks 127
    scrying 131
    werewolves 106
    witch of Scrapfaggot Green 118
Wizards 184
Woburn Abbey ghosts 58
Wooldridge, Anthony 103
Work ghosts 61–4
Worth, Patience 88
Wright, Elsie 34
Writing automatic 74, 83, 88, 143
Wyllie, Timothy 112

Xiong Jie 130

Yakut peoples familiars 107
Yang Feng Shui 55
Yarralumba House ghost 59
Yeti 102–3
Yin Feng Shui 55
York, Susannah 43–4
Yorkshire
    Arctic foxes 100
    poltergeists 47
Yugoslavia vision 52

Zapotec peoples totem animals 106
Zener cards 130, 140
Zeus 107, 170
Zink, David 76
Zodiac
    four humours 126
    Glastonbury 92
Zoologists 172

Index compiled by INDEXING SPECIALISTS, Hove,